DARWIN'S
ISLAND

DARWIN'S ISLAND

The Galapagos in the Garden of England

STEVE JONES

Little, Brown

LITTLE, BROWN

First published in Great Britain in 2009 by Little, Brown
Reprinted 2009 (twice)

A CIP catalogue record for this book
is available from the British Library.

ISBN HB 978-1-4087-0000-6
ISBN CF 978-1-4087-0001-3

Typeset in Bembo by M Rules
Printed and bound in Great Britain by
Clays Ltd, St Ives plc

Papers used by Little, Brown are natural, renewable and
recyclable products sourced from well-managed forests and certified
in accordance with the rules of the Forest Stewardship Council.

Mixed Sources
Product group from well-managed
forests and other controlled sources
www.fsc.org Cert no. SGS-COC-004081
© 1996 Forest Stewardship Council

FSC

Little, Brown
An imprint of
Little, Brown Book Group
100 Victoria Embankment
London EC4Y 0DY

An Hachette UK Company
www.hachette.co.uk

www.littlebrown.co.uk

To R. C. Simpson, who first taught me biology

CONTENTS

INTRODUCTION

THE DARWIN ARCHIPELAGO

Charles Darwin, as every schoolchild knows, saw the finches of the Galapagos in the years he spent there while employed as official naturalist on HMS *Beagle*. Each island had its own species, and Darwin soon worked out that they shared descent from a common ancestor; that they were a product of evolution. On his return to England he at once published his theory in his book *Origin of the Species*, which went on to prove that men had descended from chimpanzees. Nature, red in tooth and claw, had used the survival of the fittest to weed out the imperfect and, with *Homo sapiens* at the top of the evolutionary tree, had achieved her desired end. Racked by guilt at replacing the doctrines of the Church with a joyless vision of man as a shaven primate in an amoral universe, Charles Darwin retired into obscurity. He repented his blasphemy on his deathbed and was buried as a venerable and almost forgotten savant whose work – like that of so many famous scientists – had been completed while he was still a young man.

That is an entire parody of the truth. Darwin was not a hired biologist but paid for his own trip as gentleman-companion to the *Beagle*'s captain. He spent but five weeks of the five-year voyage in the Galapagos, with just half the time passed on shore, on only four of the dozen or so members of the group. He had little interest in his collection of finches and lumped their corpses together as a jumbled mass without even making a note of where they came from. Many of the famous birds live on several islands rather than one. Two decades passed before the publication of *The Origin of Species* (in which the word 'evolution' does not appear) and in that time its author wrote several substantial books. The phrase 'the survival of the fittest' is not his but was invented by the philosopher Herbert Spencer to summarise the notion of natural selection, the central element of evolutionary theory. The bloody fangs and fingernails of Mother Nature were themselves thought up by Tennyson a decade earlier not as a philosophy of life but in memory of the death of a friend. Evolution has no end in view and men do not descend from chimps, although the two share a common ancestor (an idea not explored by Darwin for a dozen years after *The Origin*). The Church soon accommodated his ideas, which, as most clerics realised, have no relevance to religion and the deathbed conversion is a simple falsehood, even if the great naturalist was buried in Westminster Abbey, where he still lies, trampled by tourists.

The most widespread error is to assume that the *Beagle* voyage marked the end of Charles Darwin's scientific career. In fact, in the four decades that remained to him after he came home from the wilds in 1836, Captain Fitzroy's gentleman-companion worked as hard as or harder than he had as a young man. He soon purchased Down House, south of London, in the eponymous village (whose name gained a terminal 'e' at the insistence of the Post Office, a rule that Darwin ignored).

At first he saw the place as dull and unattractive enough, but before long the house was transformed, with the help of his considerable fortune, into a grand but comfortable mansion. Its owner settled in the land of his birth and never left again: uxorious, paternal and reluctant to leave his extensive garden except on forays to test his theories and, now and again, to search for better health. As he wrote, with some satisfaction, many years after moving in: 'Few persons can have lived a more retired life than we have done . . . My life goes on like clock-work and I am fixed on the spot where I shall end it.' So settled was he that he described his profession as 'farmer' in the Bagshawe's *Directory* of the time. Great Britain was the first and last of the forty islands he visited and the patriarch of Downe studied its natural history in far more detail than he had that of anywhere else. His own county of Kent – the Garden of England – was as much, or more, a place of discovery than had been the jungles of the Amazon or the stark cinders of the Galapagos. The British Isles were where Charles Darwin built his reputation.

This book is about the disregarded Darwin, the most illustrious figure in biology, and about his years of work on the plants, animals and people that make their home in the land of his birth. *The Origin of Species* is, without doubt, the most famous book in science. It celebrates its hundred-and-fiftieth anniversary in 2009, which marks in addition the author's bicentenary.

To remember that *magnum opus* alone would be as foolish as to celebrate Shakespeare just as the author of *Hamlet*. The great naturalist's lifelong labours generated an archipelago of information; a set of connected observations that together form a harmonious whole. He wrote six million words in nineteen published works, hundreds of scientific papers and innumerable letters, fourteen thousand of which survive. Although – because of the famous note from Alfred Russel Wallace that bounced him into writing *The Origin* – he never finished his *magnum opus*, his 'big species book', much of its planned contents appeared as a series of separate volumes throughout his lifetime. Biology emerged from that gargantuan effort as a unitary subject, linked by Charles Darwin's grand idea of common ancestry, of evolution. The volumes that poured from his comfortable study were guidebooks that made sense of a whole new science. They allowed its students to navigate what had, before his day, been an uncharted labyrinth of shoals, reefs and remote islets of apparently unrelated facts.

The Origin itself was in truth no more than a prologue to the great man's career. It is as much a work of reportage as it is of research. Most of his other publications are, in contrast, based on his own observations and experiments and explore, with his trademark enthusiasm, what appear at first sight to be almost unrelated aspects of the natural world. Darwin's domestic works, as they might be called, are, in order of appearance and with titles somewhat truncated: *Barnacles* (in four volumes), *Orchids and Insects, Variation under Domestication, The Descent of Man, Expression of the Emotions, Insectivorous Plants, Climbing Plants, Cross and Self-Fertilisation, Forms of Flowers, Movement in Plants* and *Formation of Vegetable Mould by Earthworms. The Origin* has but a single illustration, but most of the others are filled with line drawings, engravings and plates, almost five hundred altogether (and some find a place in the present pages). *The Expression of the Emotions* was one of the very first scientific books to be illustrated with photographs.

His literary oeuvre was aimed at a wide audience and is set out in good, plain Victorian prose. He wrote to Thomas Henry Huxley in 1865 that 'I sometimes think that general and popular Treatises are almost as important for the progress of science as original work.' Charles Darwin was the first popular science writer – and his publisher appreciated as much for he gave *The Origin* equal billing with Samuel Smiles's quintessentially Victorian work *Self-Help*, which appeared on the same day. The author himself realised the public's interest in his work for he was one of the first among that dubious breed of scribblers to negotiate a pre-publication cash advance before settling down at his desk. Unlike most of his intellectual descendants, Darwin's command of foreign languages was good enough to allow him to pick up some of the atrocities committed on his manuscripts by their translators and he spent much time anguishing about quite what French or German phrase best approximated to his central notion of 'natural selection'.

Here I attempt to bring his lesser-known writings up to date for the modern age and to place the world's pre-eminent biologist firmly in the context of his native land. His literary canon makes sense only when considered as a whole. At first sight its subjects seem almost disconnected – earthworms, inbreeding, barnacles, plant hormones, domestication, insect-eating plants and the expressions of joy or despair in dogs, apes and men – but in truth all share a theme: the power of small means, given time, to produce gigantic ends. Fond family man as he was, he saw no gulf between the powers that had made his wife and children and those at work elsewhere. His concerns about the risks of marrying his cousin were tested with experiments on flowers. In the same way, an interest in the emotions of animals led to a comparison of the expressions of his infant son with those of dogs and apes. Different as his children might be from such humble creatures, all had emerged through the action of the same biological forces; through evolution, or 'descent with modification'. The notion, and his willingness to apply it to ourselves, outraged some of his fellows. It leaves many people uncomfortable today.

Biology has plenty of heroes but Charles Darwin is unique, for he was a pioneer in so many of its branches. He became a better scientist as he grew older for he began to test ideas with his own hands-on research, much of it far ahead of its time, rather than collating the results of others, brilliant as the synthesis might be; as he said later in life, 'I am like a gambler, & love a wild experiment.'

A good portion of the educated public has heard of *The Origin* and *The Voyage of the Beagle* but his other works are almost unknown. Most biologists are familiar with at least some of them for each volume is a milestone in their profession. The *Earthworms* epic founded modern soil science, *Emotions* saw the dawn of comparative psychology while *Cross and Self-Fertilisation* and *Forms of Flowers* were each an attempt to understand the origin of sex. The experiments described in *Movement in Plants* gave the first clue to the existence of hormones (although the word had not been invented and their discovery in animals had to wait thirty years). Their author also wrote on carnivorous plants, on the links between insects and orchids, and on the origin of our domestic plants and animals (and there he grappled with the nature of heredity, and almost got it right, with talk of crosses between round and wrinkled peas). Even his four books on barnacles, obscure as they appear, are important, for they showed that juvenile forms reveal more about relatedness than do adults and that bodies as complicated as our own are built on a simple plan. For barnacles and all other creatures his mechanism of natural selection generates organs of impressive perfection not by design but by tinkering with whatever raw material is available.

The Descent of Man, and Selection in Relation to Sex, to give its full title, stands rather apart from the rest. The book appeared in 1871 and was both the first real treatment of human evolution and an introduction to the importance of sexual conflict in evolution. It sets out the entire Darwinian argument with reference to a single group of creatures: man and his relatives. *Descent*, like *The Origin of Species*, is in the main a compilation of the results of others. Even so, it fits well into what might be called the Down House School and I use it here as an introduction to the world of modern evolutionary biology as illustrated by ourselves and our primate relatives. The study of our past has been transformed. If the author of *The Origin* were to rewrite that famous work today he would turn for many of his examples not to pigeons and tortoises, nor to worms and barnacles, but to his fellow citizens. *The Origin*'s sole mention of *Homo sapiens*, the tentative claim that 'Light will be thrown on the origin of man and his history', has been wonderfully upheld. It shows how the truths glimpsed by Darwin now unite the whole science of life.

Here I attempt to update all those topics for today. I append an *envoi* with a look at the biological world of the twenty-first century compared with its state in 1859. In the tradition of the great naturalist himself, who was dubious about the many infantile attempts to apply his ideas to society (as in a newspaper's claim that his work proved that 'might is right & therefore that Napoleon is right & every cheating tradesman is also right'), I avoid as far as possible any discussion of the relevance of Darwinism to the human predicament. I also steer clear of the empty arguments about its interactions with religion. The struggle to separate science from theology still fascinates a few, but most scientists have no interest in it (although there are exceptions, for the Victorian biologist Thomas Henry Huxley felt that 'Extinguished theologians lie about the cradle of every science as the strangled snakes besides that of Hercules'). Today's biology in its success emphasises how little relevance it has to the issues so often, and so tediously, discussed by non-biologists. As Darwin put it in *The Descent of Man*: 'We are not here concerned with hopes or fears, only with the truth as far as our reason allows us to discover it.' Science can do that, and no more.

My eminent predecessor at University College London, the Nobel Prize-winner Peter Medawar, in an acerbic comment on the relative merits of students of science and the arts, said of Watson and Crick (of double helix fame) that 'Not only were they clever, they had something to be clever about.' Not only did Charles Darwin travel, he had something to travel for. The joy of the *Beagle* voyage was that it had a point. For a real adventurer, to travel hopefully is not enough: some end must be in view. As he wrote in the last pages of his account of the journey: 'If a person asked my advice before undertaking a long voyage, my answer would depend upon his possessing a decided taste for some branch of knowledge, which could by this means be advanced.'

Darwin's odysseys, from the Galapagos to West Wales, play an important part in all his books, as they did in the author's life. The *Beagle* crossed nearly fifty thousand kilometres of ocean but his British journeys covered almost as much country. His work was always tied to where he found himself, whether in a rain forest or a suburb. Many of his compositions emerge from a kind of Grand Tour of the British Isles. His very first memory, as recounted in his autobiography, was of a visit to Abergele for the sea-bathing at the age of four. Six years later he was back on the Welsh coast at Towyn, where he noted some 'curious insects' (black and red Burnet Moths) not seen around Shrewsbury. Unlike the many naturalists of those times who filled cabinets with butterflies or shells to make a biological stamp-collection, he wondered, even as a child, quite why they were found in one place and not another.

As he grew older, natural history became an all-embracing passion. His early enjoyment of literature, art and music disappeared and he wrote that 'I have tried lately to read Shakespeare and found it so intolerably dull that it nauseated me.' His preferred reading consisted of romantic novels (the sillier the story, the better, said his children) and he sold off his family heirlooms of Wedgwood pottery and Flaxman reliefs. He could make out 'absolutely nothing' of what merit there was in a collection of Turner watercolours. 'My mind', he wrote, 'seems to have become a kind of machine for grinding general laws out of a large collection of facts, but why this should have caused the atrophy of that part of the brain alone, on which the higher tastes depend I cannot conceive' (although he did send the *Expression of the Emotions* book to art journals for review, where it was criticised for its insensitivity to the nature of Art).

That obsession with science allowed Charles Darwin's juvenile interest in the insects of England and Wales to grow into a lifelong exploration of the biology and geology of his native island. He published his first scientific paper, on the eggs of an animal found in the Firth of Forth, in the twelve grey months he spent in Edinburgh. After a brief visit to Dublin, the young enthusiast then moved to Cambridge, where he spent many days knee-deep in bogs and fens in the search for specimens. Just before the departure of the *Beagle*, he travelled for three weeks across North Wales from Shrewsbury to Conwy and Barmouth with the geologist Adam Sedgwick, who taught him the elements of mapping so useful on the voyage. On his return he set off again to Scotland, where, in his first major scientific paper, he made a frightful error in his evaluation of a series of parallel shelves or 'roads' in Glen Roy as wave-cut beaches rather than the shores of drained glacial lakes (as he wrote many years later, 'I am ashamed of it'). Later in life he criss-crossed Britain to pursue his researches or to take his family on holiday, or to escape the epidemics of infection that now and again swept through Downe (and killed two of his own children). They went to Wales, to the Isle of Wight (where he met Alfred, Lord Tennyson), to Torquay, to the Lake District (an audience there with Ruskin), to Stonehenge, to the heathlands of England and to a variety of grand mansions across the kingdom. Often, his experimental subjects – pots of orchids or of insect-eating plants – travelled with the family, at considerable inconvenience. He had plenty of time to explore the British Isles for in his forty years at Down House Charles Darwin spent two thousand nights away from home – the equivalent of a day a week. A few of his trips lasted a month and more.

Some of his travels were in search of science, but many were a quest for health. He became chronically ill very soon after his return from the *Beagle* trip and his heavy use of snuff and tobacco did nothing to improve his well-being. Darwin visited spas in Great Malvern, in Guildford and in Ilkley (where he received the first copy of *The Origin*). His later years were marked by a series of bizarre attempts to remedy his feeble state (even if he did write that illness, 'though it has annihilated several years of my life, has saved me from the distractions of society and amusement'). The main symptom was vomiting, often brought on by stress, with the rushed last chapter of *The Origin* sparking off a severe episode that caused great prostration of mind and body. So severe were the attacks that he declined some invitations to stay in friends' houses on the grounds that 'my retching is apt to be extremely loud'.

He tried Condy's Ozonised Fluid, 'enormous quantities of chalk, magnesia & carb of ammonia', and rubber bags filled with ice and worn next to the spine. Nothing worked (although he learned to play billiards at one of the establishments and became a devotee of the pastime, which helped him to relax and, as he said, 'drives the horrid species out of my head'). The author of *The Origin* was a victim of the Victorian 'Demon of Dyspepsia' and was joined in that unhappy throng by Thomas Carlyle, George Eliot, Charles Dickens, Florence Nightingale and the evolutionists T. H. Huxley, Alfred Russel Wallace and Herbert Spencer, together with his own brother Erasmus. Their troubles funded several pharmaceutical fortunes (including that of Henry Wellcome, which later helped pay for that Darwinian triumph, the sequence of human DNA). What his condition might have been is not known: a supposed conflict between Christian belief and rationalism, or a parasite picked up in Brazil or even, some say, the obsessive swallowing of air. He was diagnosed as having 'waterbrash' – heartburn, in modern parlance, the reflux of acid from the stomach – which can result from an ulcer. Dyspepsia's nausea, depression and lassitude are, we know today, caused by a bacterium. The bug that swept through Victoria's intellectuals might now be cured with a simple pill.

Later in life, in part because of his health, the paterfamilias of Down House spent longer and longer periods without leaving home. He fed his household with fifty-three distinct varieties of gooseberries and three of cabbage. In his garden he carried out many experiments, helped by William Brooke, his 'gloomy gardener' (who was seen to laugh just once, when a boomerang broke a cucumber frame). The naturalist's tale ends, in the tradition of the classics, with the hero's death and his desire to join his beloved earthworms in the 'sweetest place on Earth', the village churchyard at Downe – a wish frustrated by fame, the establishment and the Abbey.

Darwin's Island retraces some of Darwin's steps and moves his discoveries forward by a century and more. It will, I hope, help bring his less well-known work into the third millennium. Several people have helped in the preparation of this book. David Leibel, Michael Morgan, Kay Taylor and Anna Trench made helpful comments on parts of it. I thank them for their help.

Three of my earlier volumes – on coral reefs, on the nature of maleness and on the theory of evolution itself – pay homage to the founder of the science of life, and each is an attempt to update his ideas for the modern age. There could be no better way to honour the most famous of all biologists at this time of concentrated attention on his history than to give his less celebrated works the exposure they deserve. For Charles Darwin, the five *Beagle* years that became part of Britain's intellectual legacy led to four decades of intense labour within the confines of his native land. In that modest group of islands he underwent a second great voyage: not of the body but of the mind. This book traces that journey from its beginning to its end.

CHAPTER I

THE QUEEN'S ORANG-UTAN

In 1842 Queen Victoria went to London Zoo. She was less than amused: 'The Orang Outang is too wonderful . . . he is frightfully, and painfully, and disagreeably human.' The animal was not a male but a female called Jenny and Charles Darwin had, some years earlier, visited its mother. He too spotted the resemblance between the apes on either side of the bars. The young biologist scribbled in his notebook that 'Man in his arrogance thinks himself a great work. More humble and I believe true to consider him created from animals.' Seventeen years after Victoria's visit, in 1859, he published the theory that proved the Queen's kinship, and his own, to Jenny, to every inmate of the Zoological Gardens and to all the inhabitants of the Earth.

The Origin of Species caused uproar among the Empress of India's subjects. Her Chancellor, Benjamin Disraeli, asked famously: 'Is man an ape or an angel? My Lord, I am on the side of the angels.

I repudiate with indignation and abhorrence these new fangled theories.' Many of his fellow citizens agreed. Even so, the notion at once entered public discourse (and *Punch* devoted its 1861 Christmas annual to gorilla-like humans and their opposites). In time, and with some reluctance, the notion that every Briton, high or low, shared descent with the rest of the world was accepted. A quarter of a century on, W. S. Gilbert penned the deathless line that 'Darwinian man, though well behaved, at best is only a monkey shaved' and the idea of *Homo sapiens* as a depilated ape became part of popular culture, where it belongs. Victoria herself congratulated one of her daughters, the crown princess of Prussia, for turning to *The Origin*: 'How many interesting, difficult books you read. It would and will please beloved Papa.'

As the Queen had noticed, the physical similarity of men to apes is clear. In 1859, Charles Darwin came up with the reason why. A certain caution was needed to promote the idea that what had made animals had also produced men and women, and he waited for twelve years before he expanded on the subject. *The Descent of Man* describes how – and why – *Homo sapiens* shares its nature with other primates. The book uses our own species as an exemplar of evolution.

To the founder of modern biology, man obeyed the same evolutionary rules as all his kin and shared much of his physical being with them; as the book says, in its final paragraph, he still bears 'the indelible stamp of his lowly origin'. In moral terms *Homo sapiens* was something more: '. . . of all the differences between man and the lower animals, the moral sense or conscience is by far the most important. This sense . . . is summed up in that short but imperious word "ought," so full of high significance. It is the most noble of all the attributes of man, leading him without a moment's hesitation to risk his life for that of a fellow-creature.' No ape understands the meaning of 'ought', a word pregnant with notions quite alien to every species apart from one. Even so, despite that essential and uniquely human attribute, every ape –

and we are among them – is, like every other creature, the product of a common biological mechanism.

The logic of evolution is simple. There exists, within all plants
and animals, variation passed from one generation to the next.
More individuals are born than can live or breed. As a result, there
develops a struggle to stay alive and to find a mate. In that battle,
those who bear certain variants prevail over others less well endowed.
Such inherited differences in the ability to transmit genes – natural
selection, as Darwin called it – mean that the advantageous forms
become more common as the generations succeed each other. In
time, as new versions accumulate, a lineage may change so much
that it can no longer exchange genes with those that were once its
kin. A new species is born.

Natural selection is a factory that makes almost impossible
things. It manufactures what seems like design with no need for a
designer. Evolution builds complicated organs like the eye, the ear
or the elbow by piecing together favoured variants as they arise.
Almost as an afterthought, it generates new forms of life.

Its tale as told in *The Origin of Species* turns on the efforts of
farmers as they develop new breeds from old, on changes in wild
creatures exposed to the rigours of nature and the demands of the
opposite sex, on the tendency of isolated places to evolve unique
forms, and on the silent words of the fossils that tell of a planet as
it was before evolution moved on. Its pages speak of the embryo
as a key to the past and of how structures no longer of value
and others that appear almost too perfect are each testimony of
the power of natural selection. The geography of life, on islands,
continents and mountains, is also evidence of the common
descent of mushrooms, mice and men. Most of all, life's diversity
can be arranged into a series of groups arranged within groups, of
ever-decreasing affinity, as a strong hint that they split apart from
each other longer and longer ago.

The Descent of Man uses that logic to disentangle the history of
a single species. Unique as it might think itself, *Homo sapiens* is

animal like all others. The book's famous last sentence reads, in full: 'I have given the evidence to the best of my ability: and we must acknowledge, as it seems to me, that man with all his noble qualities, with sympathy which feels for the most debased, with benevolence which extends not only to other men but to the humblest living creature, with his god-like intellect which has penetrated into the movements and constitution of the solar system − with all these exalted powers − Man still bears in his bodily frame the indelible stamp of his lowly origin.'

In 1871 − and even in 1971 − the evidence for that final and provocative statement was weak indeed. Now, everything has changed. The entire evolutionary case can be made in terms of ourselves and our relatives; of apes and monkeys, of chimps and gorillas, and of men and women. Our new ability to look at genes, cells, tissues and organs in exquisite detail means that we know more about the human past than about that of any other species. Evolution is best viewed through our own eyes; and not just because we are all interested in where we came from but because advances in science mean that *Homo sapiens* has become the embodiment of every evolutionary idea. Darwin's theory has not altered much in the century and a half since it was proposed. The technology used to study it has, on the other hand, been transformed.

Technical as they have become, the tools used today to examine the past would have been familiar in their nature, if not in their details, to biologists of the nineteenth century. Charles Darwin was, among his many talents, a proficient anatomist. He used changes in the physical structure of pigeons, pigs and people as evidence for his theory. The first chapter of *The Descent of Man* is a somewhat ponderous account of the differences between the bones and bodies of men and apes. Dissection, once at the centre of biology (and biologists of a certain age still flinch at the smell of formalin), not long ago appeared antiquated, but now it looks very modern. Molecular biology is no more than comparative anatomy plus a mountain of cash. Its chemical scalpels cut up creatures into thousands of millions of individual letters of DNA code. Those who wield them have shown beyond all doubt the truth behind Queen Victoria's fear that the bodily frame of Jenny the orang-utan was proof of the common ancestry of humans with apes and with far more.

The Human Genome Project – the scheme to read off our own DNA sequence – set the seal on an enterprise which began in the sixteenth century when Vesalius opened the heart and discovered that it had four chambers rather than the three insisted on by the Greeks. Its completion was announced in 2000 and again in 2003, 2006 and 2008 (and some parts of the double helix still remain unread). A science that had been in its infancy a mere description of bones and muscles became an adolescent when *The Origin of Species* showed how shared structure was evidence of common descent. It has at last matured. The anatomy of DNA has become the key to the history of life.

In a glass-fronted cabinet at University College London resides the stuffed body of the eighteenth-century philosopher Jeremy Bentham, the 'greatest good for the greatest number' man. His Auto-Icon, as he called it, was an attempt at a memorial that would cost less than the showy shrines then fashionable. Bentham was convinced that his idea would catch on. Two centuries later, it did. James Watson – the surviving half of the duo who unwound the double helix – was presented with his own auto-icon, a compact disc of his entire DNA sequence, which he can, if he wishes, display for public edification in a small plastic case.

Watson's essence is coded into a tangled mass of intricate chemistry. The egg that made him contained two metres of DNA and each of the billions of cells that descend from it as his body grows and ages has a copy. Each of those molecular sentences is written in three thousand two hundred million letters, the four bases of the familiar genetic code. Twenty years ago, when the scheme to read the whole lot off was proposed, it took months to decipher the number of letters found in this paragraph. The molecule was sliced into random bits, each was read from end to end and the whole genome stitched together with a search for places where the fragments overlap. Such methods are antique. Today's machines pick up flashes of light from molecules tagged with fluorescent dyes, each base with its own colour, and squeezed one at a time through tiny pores. It takes no more than a few hours to read off a piece as long as this entire book, which itself contains less than one part in several thousand of the whole content of the human genome. Soon it will become possible to sequence single molecules rather than multiple copies, as is now necessary, and enthusiasts speak of machines that will read off a million DNA bases a second.

The first human sequence cost up to a billion dollars and Watson's version was auctioned off for a million. In 2008 the Knome Corporation offered to read off the DNA of anybody with a spare $350,000. In fact, the whole lot can now be done for a fraction of that sum. Within five years the price will drop to a few thousand dollars per genome and it will become possible to decipher the DNA of any creature at nominal cost. The web of kinship that binds life together will then be revealed in all its details.

The raw material of evolution is, in its physical structure as an intertwined helix, simple or even elegant, but in its biology is entirely the opposite. In its details DNA is, frankly, a mess, for natural selection has been forced to build upon what it already has. Life did not emerge from engineering, but from expedience. The Darwinian machine has no strategy and can never look forward. Its tactics are based on the moment, and the genomes it makes, like the creatures they code for, are the products of a set of short-term fixes. James Watson's molecule is marked by redundancy, decay and the scars of battles long gone. Genes – like cells, guts and brains – work, but only just.

Human DNA contains long stretches that appear to be useless

and numerous sections that are mirror-images of each other. Repetition is everywhere: of particular genes into families that carry out similar tasks and of multiplied lengths of material that seems redundant. The remnants of viruses make up almost half the total and the remainder is littered by the decayed hulks of ancient and once functional structures. All but one part in fifty of the genome was, as a result, once (mistakenly) dismissed as biological garbage.

The genes themselves have become blurred and ambiguous as we learn more. There are far fewer than expected when the genome project was proposed – just over twenty thousand rather than the multitude then assumed to be essential. Some overlap with each other or say different things when read in opposite directions or when active in different tissues. Many contain inserted sequences of DNA that looks as if they have no function (although some of the supposed junk does a useful job while other sections cause disease should they wake up and shift position). Plenty of questions remain. How important is the part – often a small part – of each gene that codes for proteins compared with the on and off switches, the accelerators and brakes, and the rest of the control machinery? We do not know.

Even the size of the package makes little sense. A chicken has slightly less DNA than does a Nobel laureate but half its genes are identical, or almost so, to our own – evidence, given that we last shared an ancestor three hundred million years ago, of how conservative evolution can be. A tiny plant called *Arabidopsis*, a relative of the Brussels sprout, has more genes than either. All this says more about how hard it can be to define what a gene actually is than about the talents of sprout versus sentient being.

Eight decades passed between Vesalius' dissection of the heart and the discovery of the circulation of the blood. The genome is now in that transitional period. DNA's nuts and bolts (and even some of its bells and whistles) have been dismantled, but most of those who work on it still study structure without much insight into function. William Harvey (the circulation man) saw the heart as a mere pump, and understood nothing of its exquisite system of control. Genes are much the same. Each is linked into a network with others and responds to messages from both within and outside the cell. The path from instruction to product is a labyrinth, rather than a straight line. The proteins that pour from the cell's biological factories are not simple blocks that slot together but are folded, spliced, cut, or fused into new mixtures in a way that depends on local conditions almost as much as upon their own structure. Diseases as different as diabetes and prostate cancer may arise from damage in the same segment of DNA, while others such as breast cancer emerge from errors in several different genes. Most of the double helix is switched off the majority of the time, African genes are, on average, more active than are those of Europeans and life has begun to look far more complicated than any molecular biologist had feared.

Evolutionists are not in the least surprised. They were baffled at some of the decisions made by those who ran the Genome Project. Like Vesalius, James Watson and his colleagues had a Platonic view of existence. Every heart and every human was built on the same plan and to understand one was to understand them all. The first DNA sequencers outPlato-ed Plato for they assumed not just that the essence of humankind could be found within a single person, but that this Mr Average was, in the interests of political correctness, best stitched together from bits of double helix taken from random donors across the globe.

That was a big mistake. The Platonic approach ignores the vital truth that evolution is a comparative science. Natural selection depends on inherited differences. To understand the past biology needs not just a single genome but many. To map variation from person to person, from place to place, or from species to species shows how, when and where evolution has been at work. So central is diversity to the idea of descent with modification that the first two chapters of *The Origin* are devoted to the nature and extent of variability in the bodies and habits of plants and animals. Now, genetics has begun to tell the tale in the language of DNA.

James Watson's auto-icon disclosed no more than half his secrets for it contained just one of the two versions of the double helix present in each cell. His rival in the race to decipher the secrets of life, the biologist and businessman Craig Venter, was less reticent. He read off both his copies, that received from his father and that from his mother. Venter was happy to reveal its contents: his father died young of a heart attack, and he has himself been bequeathed a variant that predisposes to the disease. He has also inherited genes supposed to increase the wish to seek novelty, to be active in the evening rather than early in the day and to have wet rather than dry ear wax.

Whatever Venter's intimate chemistry says about his personality, his bed-time or the exudations of his auditory canal, it has a message for us all for it gave the first hint of the true level of human diversity. Both his parents are white Europeans (and hence represent just a small sample of mankind) but their DNA is distinct at around one site in two hundred along the entire chain – which adds up to tens of millions of differences between them.

On the global scale, hundreds of millions of sites in the inherited alphabet vary from person to person and the 'Thousand Genomes Project', now well under way, has set out to fund out just how many there might be. Unlike its predecessors it will search out rare variants, those carried by fewer than one person in a hundred and present in vast abundance – and given the advances of technology, the project may cost little more than fifty million dollars. Already we know that each of the twenty-three human chromosomes – the physical location of the genes – has millions of single-letter changes aligned along it. The variable sites are so tightly packed together that, over short lengths of the double helix, they almost never separate when the molecule is cut, spliced and reordered, as it always is when sperm or egg is formed. Such long blocks represent sets of chemical letters that travel down the generations together. Rather like surnames, they are excellent clues of relatedness.

In addition to its single-letter changes, the double helix is marked by duplications of certain pieces and deletions of others. The order of its letters may also be reversed, and great stretches can hop to a new place. A study of three hundred whole genomes has already revealed a thousand and more such differences in the numbers of particular DNA sequences. Some genes are arranged in families – groups of similar structures that descend from a common ancestor and have taken up a series of related jobs. The biggest has eight hundred members. It helps build the senses of taste and smell. Its elements vary in number from person to person and some lucky individuals have fifty more copies of a certain scent receptor than do others.

Most such changes involve fewer than ten letters, but some are a million base-pairs from end to end. A few people may, because of the gains and losses, have millions more DNA bases and thousands more genes than do others and the potential variation in dose from person to person represents more than the length of the largest human chromosome. Even so, some of the repeated segments have just the same structure in humans as in the coelacanth, which split apart four hundred million years ago.

DNA is a labile and uncertain molecule. A multiplied sequence often makes mistakes as egg or sperm are formed, to produce longer or shorter versions of what went before. Some bits move or multiply at a rate of one in a hundred each generation rather than the one in a million once assumed to be typical. Age changes us and the double helix is reordered, duplicated and deleted as the years go by (which means that the offspring of older parents inherit more mutations than do those of young).

Variability beneath the skin is far more extensive than Darwin had ever imagined. Biologists have long known that, with the exception of identical twins, everyone in the world is distinct from everyone else, and from all those who have ever lived, or ever will. That claim is too modest. In fact, every sperm and every egg ever made by all the billions of men and women who have walked the Earth since our species began is unique; a figure unimaginable before the days of molecular biology.

Such variety links individuals, families and peoples into a shared network of descent. It shows how man is related to chimpanzees, gorillas, orangs and macaques, and for that matter to plants and to bacteria. Evolution – like astronomy – has always looked at the past through the eyes of the present but its new technology – like the star-gazers' development of giant telescopes – means that it can now see far further and deeper into the universe of life than once it could.

Even so, biology is not like astronomy. The images that flood from its machines are often blurred and ambivalent. Many statements about ancestry are filled with unproven, and often unstated, assumptions about the rate of change in DNA, the size of ancient populations and the effect – or supposed lack of effect – of mutations on the well-being of those who bear them. The information in the genome is almost limitless, but at present its language remains ambiguous.

Fortunately, the Earth has some better witnesses to years gone by. Like the remnants of stellar rocks that sometimes strike our planet, they are silent, shattered and few in number but at least they give direct evidence of how the past unfolded. Darwin was well aware of the importance of the fossil record to his case. One page in six of *The Origin* is devoted to the relics of the rocks, to the record's imperfections and to the central role it plays as proof of the fact of change. In 1871, no human fossils (with the exception of a skull from Germany now known to come from a Neanderthal) had been recognised. Things have much improved and the primate record is far more complete than it was even a few decades ago. The tale it tells is still fragmented and uncertain, but what it says fits remarkably well with the history revealed by the double helix.

In the Miocene epoch – from around twenty-three million to five million years ago – the Earth was a true Planet of the Apes. Primates were all over the place, with a hundred or more distinct species of ape, in Africa, in Asia and in Europe. They lived in woodlands, plains, forests and swamps. Some were no bigger than a cat and others larger than a gorilla. For much of the time their capital was in Europe and many of our predecessors have laid their bones there. Then the animals moved on, to set up shop in Africa. A ten-million-year-old fossil from Kenya may be the common ancestor of men, chimps and gorillas. If so, it confirms Darwin's speculation that it was more probable 'that our early progenitors lived on the African continent than elsewhere'. He did not, of course, know that continents had broken up and drifted across the world, and that Africa itself did not exist in the earliest days of the evolution of our line.

One day almost all the players in that ancient drama left the stage. The apogee of the apes was over and their long twilight – now fast turning into night – had begun. The sun began to set on their family well before humans appeared, but, once they did, their nemesis was assured.

Lucy, the famous fossil of *Australopithecus afarensis*, was a creature quite human in appearance, lightly built and little more than a metre tall, with relatively long legs and small teeth. She belonged to a group who lived between three and four million years before the present. Others among her kin left footprints in Tanzanian volcanic ash as proof that they walked upright at a time when their brains were but a third the size of our own. The males were considerably larger than the females. *Homo habilis* – 'handy man' – lived in South and East Africa for about a million years from two and a half million years ago. It had long arms, brow ridges and a larger brain than Lucy, and was quite good at making tools. Similar individuals were found in Africa, and perhaps in Georgia.

Homo erectus, the upright human, the next fossil claimed as a direct (or almost direct) human ancestor, emerged around 1.8 million years ago, and may have split into two species in its homelands in Africa and Asia. Some individuals had brains as large as our own and lived as far north as the South of France. A rather younger European arrived around 1.2 million years before the present, and left a few of his bones in the caves of the Sierra de Atapuerca in northern Spain. That ancient Spaniard has been christened *Homo antecessor*, and might be the common ancestor of ourselves and the Neanderthals. A later European from around half a million years ago, Heidelberg Man, may have been an antecedent of the Neanderthals rather than ourselves. He too first appeared in Africa. Many – perhaps too many – more supposed members of our close family have been named as distinct species, and the human pedigree has begun to look more like a bush than a tree. As a result, direct lines of descent have become harder to trace than once they were.

For most of history, our ancestors shared their home with several related species that were much closer to themselves than the chimpanzee is to us. Those days have gone, and nearly all members of man's ancient household have left no issue today.

The Neanderthals were once our most immediate kin. They lived in Europe and the Middle East from around a quarter of a million years ago to about thirty thousand. They had bigger skulls – and, perhaps, bigger brains – than modern humans (although they were also beefier in general). They trapped animals in pits, and may have been cannibals (although another view of the carved bones of their fellows is that they represent a ritual burial). Neanderthals lived in small groups in an icy Europe for far longer than our own species has existed, and then disappeared. Like many other apes, they went quickly. Perhaps a cold snap defeated them, for a remnant hung on in the warmth of southern Spain until well after the moderns arrived. The victors had better clothes, which allowed the tropical ape that they were – and we are – to survive in a climate that killed off an animal more used to bad weather but less well clad. Perhaps *Homo sapiens* murdered the Neanderthals or starved them out, but we do not know. Sex was not on the agenda, for fossil DNA from a Croatian specimen shows that they were quite distinct from our direct ancestors. In addition, today's Europeans and Middle Easterners retain no ancient lineages that might have come from an extinct relative. DNA suggests that the Neanderthals' last common ancestor with modern humans lived in Africa more than six hundred thousand years ago, long before *Homo sapiens* emerged.

Soon after the loss of his cousin, that species began to spread across the world. Modern humans filled the whole habitable globe no more than a thousand or so years before the present, when at last men and women reached New Zealand and Hawaii. Their ancient journeys can still be read in DNA. The double helix reveals a clear split between Africa and everywhere else, a legacy of the small group of migrants who first stepped out of our native continent into an uninhabited world, together with a second and more ancient split within Africa that separates the Khoi-San – the Bushmen – from all others. Other great genetic trends, such as those across the New World and the Pacific, track the last migrations into a deserted landscape.

Once, it seemed that modern Europe had a more complicated history than did most of the globe, with several waves of migration superimposed on each other. The genes of local hunters, who arrived long ago, were – perhaps – diluted by those of the first farmers who spread, just a few thousand years before the present, from a population explosion in the Middle East. Some variants do show a trend from south-east to north-west, in a pattern that might indeed reflect a slow wave of inter-communal sex. The archaeology of pots and seeds suggests in contrast that agriculture was taken up at some speed, as soon as people learned about it, with no need for weddings. In Britain, at the western edge of the new technology, carbon dates taken from charred grains suggest that around 4000 BC farming replaced hunting within just a couple of centuries, too fast for any large-scale mixture of populations. There is no real evidence of a flood of lascivious rustics coming from the east. Instead, ancient Europe was more open to ideas than it was to genes. The trends seen today are the remnants of the first grand migration thirty thousand years before the emergence of agriculture, as humans arrived in an empty continent from the south and east. The mitochondrial DNA – the female lineages – found in the remains of a hunter-gatherer group in northern Spain look more or less the same as those of modern Spaniards in the same place, with no sign of mass immigration. Modern Europeans trace most of their heritage to the first wave of hunters. Since then, they – and their DNA – have tended to stay at home.

As men and women filled the world they killed off many of their kin. The Neanderthals were the first to go, and human habits have not changed since then. Today, just a few remnants of our once extensive clan linger on. In a century or so we will be the single large primate (and almost the only large mammal), to be found outside farms or zoos. Almost all the apes will be gone, some before they are studied by science. That fact is a tragedy both for the creatures involved and for science itself, for each of them says something about our own biological heritage. They contain within their DNA the story of human evolution and, perhaps, more: for some of our own inborn diseases are caused by genes identical to some that function perfectly well in our relatives.

The physical similarity of primates and humans was noticed by Queen Victoria and, after *The Origin*, was often used by those anxious to judge the evolutionary status of their fellow men. Charles Kingsley, author of *The Water Babies*, wrote to his wife about an Irish visit that 'I am haunted by the human chimpanzees I saw . . . to see white chimpanzees is dreadful; if they were black, one would not feel it so much.' Chimpanzees are, indeed, our closest relative. Darwin himself noted that, among their many other affinities to humans, they 'have a strong taste for tea, coffee, and spirituous liquors: they will also, as I have myself seen, smoke tobacco with pleasure'.

Whatever our shared vices, chimps are not like us in many ways. They are hairy and bad-tempered and do not show the whites of their eyes. The animals have rather small brains, no ear lobes and cannot walk upright, float, or cry when upset. They give birth with less pain than we do, and the young mature without any obvious period of adolescence. Our kidneys keep salt better in the body than do theirs, and we have more white blood cells. Chimpanzees are in addition safe from the horrors of old age as they tend to die young and even in zoos do not get Alzheimer's disease. When they are faced with diseases brought on by infection or poor diet, their symptoms often differ from our own, which means that they have not been as useful in medical research as might be hoped.

Chimp sex life has a definite flavour of its own. Men lack the penile bone found in male chimpanzees, but when it comes to penis size, man stands alone. Women have outer labia, absent in their closest relative. Chimpanzee males have larger testes than we do in relation to their body size and, unlike ourselves, seal up their mates with a sticky plug after sex. Promiscuity is the rule. The creatures copulate with enthusiasm and their close kin the pygmy chimps or bonobos are even more energetic. The females show when they are fertile (unlike women, who conceal all signs of that crucial moment) and the males then indulge in a competitive frenzy to mate with them. Sperm from rhesus macaques, a species known to be highly promiscuous, swim faster and lash harder than those of gorillas, in which a single male more or less monopolises the females. Chimpanzee sperm are almost as energetic as those of the macaque while ours lag well behind either. They do, on the other hand, beat the male cells of the gorilla.

The chimpanzee genome was read off in 2005. Not many of the single letters in the DNA code have changed since the split from our own family line for, on that simple measure, humans and their closest relative are almost 99 per cent the same. At the protein level, too, we are close, with no more than about one amino acid in a hundred having altered.

Such figures underrate the real divergence of the two species. Changes in the number and position of inserted, repeated or deleted segments mark both lines. There are three times as many alterations of this kind as of single-base changes, which puts the overall difference between men and chimps at around 4 per cent. Primates go in for the gain and loss of genes more than do other mammals and our own lineage is out in front with a rate of change three times faster than average. *Homo sapiens* has gained seven hundred gene copies since the split with chimps, and the chimp has lost almost the same number. One chromosome has gone even further. Women have two large X chromosomes, men an X matched with a smaller Y. The human and chimp X have diverged by just half a per cent in the single letters of the DNA alphabet while the Y has shifted three times more, as proof that women, with two Xs, are closer to chimpanzees than are men.

Many of the differences between the two primates have built up because we can modify our environment in a way that other primates cannot. As a result, we depend less on changes in our DNA than once we did and so have lost many once-functional genes. Mankind is feebler than it was. We became shaved monkeys with just a single mutation or a few because a segment of DNA that codes for the hair protein no longer works. It received its fatal blow a quarter of a million years ago. Samson lost his strength with his locks, and so did his ancestors, for the DNA behind certain powerful muscles is out of action in humans compared with their closest living relatives (which is why to wrestle with a chimpanzee is a mistake). A shared *déjeuner sur l'herbe* is also best avoided, for the animals have enzymes that break down poisons fatal to ourselves. Darwin noted that 'savages' ate many foods that were harmful unless cooked; and red kidney beans still fall into that category. Chimpanzees need no kitchens, for they can manage a variety of noxious plants (certain herbs used for medical treatment included) that we cannot. They have also kept many of the talents of taste and smell lost in humans, perhaps because they need to be more careful in their choice of food before they chew it. Many of our own gustatory sensors live on just as battered remnants of once-useful structures.

As well as the differences between chimpanzees and humans, each varies to some degree from place to place. The chimpanzee is strongly subdivided at the DNA level. It has three distinct 'races', in West, in Central and in Eastern Africa. The central group is about three times as diverse as is the western. The extent and pattern of diversity hints that the western and central groups split half a million years ago, while the eastern segment found an identity no more than fifty thousand years before the present. Humans are in comparison tedious, with far less change among the world's populations than among the chimp races.

Sequencing machines are now hard at work on more of our relatives. Rhesus macaques are small monkeys common in India, Burma and elsewhere in the Far East, and widely used in medical research. They share rather more than nine-tenths of their DNA with humans. Many of the shifts involve – as in the chimpanzee – changes in the order or numbers of copies of particular segments. The animals eat lots of fruit, and genes that help digest sugar have been multiplied compared with our own. Some of their genes are in a form that in humans leads to disaster. The mutation for the rare inborn disease known as phenylketonuria – a fatal inability to deal with certain foods – is the standard version found in macaques. Might some dietary change have rendered lethal to ourselves an enzyme once useful in our ancestors? A group of genes that predisposes to cancer in humans helps make sperm in macaques. Why does a sexual helper in monkeys cause our cells to run out of control? If we knew, we might have a new weapon against the disease.

Men and chimps, and men and macaques, have changed a lot since their paths parted. The fossils and the genes combine to say when and how their evolutionary divergence, and others from long before, took place. The daring assumption that DNA accumulates error at a regular rate, combined with information on dates from the scattered fossils of our distant ancestors, hints that the first true mammals evolved around a hundred and twenty-five million years ago. The double helix also shows that chimps, orangs, humans and monkeys cluster together in a class that includes lemurs and rabbits but does not admit horses, dogs, bats and many other hairy creatures. The kinship of men, lemurs, rabbits and the rest is revealed by a certain piece of mobile DNA that hops around the genome. It has been inserted in the same place in all those creatures, proof that they share a common ancestor distinct from that of their furry fellows.

Genes, fossils and geography combine to suggest that the primates as a group began around eighty-five million years before the present. The monkeys and apes split not long after that – which means that their true origin was on the vast continent of Gondwana rather than on the fragments that we now recognise as Africa, Madagascar and India. The macaque set off on its own pathway around twenty-five million years before today. The split between ourselves and our close relatives is, it appears, quite recent. The limited genetic divergence between chimpanzees and humans suggests that they separated five to seven million years ago. Their common ancestor broke away from the gorilla line a million years or more earlier, and that trio split from the orang-utan branch about six million years before that.

There is more to evolution than the random accumulation of mistakes. Darwin's machine may not have a long-term direction, but it can swerve around obstacles as they arise. At the wheel is natural selection: inherited differences in the ability to pass on genes. In its long and arduous journey, selection's ability to cope with whatever turns up has led to the physical differences between men and chimps, men and macaques and, for that matter, men and rabbits or bats. Each diverged from the same shared ancestor, and each has faced its own challenges and, with the help of natural selection, found its own unique set of solutions.

Not all of us leave descendants, but we all have ancestors. To transmit its DNA to the present day, each of them had to survive, find a mate and produce offspring. An infinity of their contemporaries tripped at one or other of life's hurdles and left no posterity. *The Descent of Man* speculates about how selection might have acted upon the human line and that of our relatives but it offered little direct evidence of its action. Sexual choice was, its author thought, important (and he began but abandoned a project to discover whether blondes were less likely to marry than were brunettes) but his case for its role was far weaker than that for animals and plants presented in *The Origin of Species*.

The evidence for lust as an engine of human evolution is still patchy at best, but that for other forms of natural selection is now compelling. The process has been – and still is – at work in our own lineage. It leaves its footprints upon the genome in many ways, some obvious and some less so. Sometimes, natural selection can be seen in action. More often, the evidence of its labours is indirect and gives no hint of how and why it was busy.

Long-term trends such as the increase in human brain size over millions of years show what selection can achieve, given time. The grand patterns of genes across the globe are also proof of its powers.

None are grander than the shifts in man's physical appearance from place to place, which are more marked than those in any other large mammal. The story of how the trends in human hair and skin colour evolved has emerged, albeit in several shades of grey, as evidence of how selection causes change and of how subtle and unexpected its actions can be.

Homo sapiens and his immediate ancestors moved not long ago from white to black, and in some places back to white again. Chimps have rather pale skins, although their faces may become tanned. African skin is, in contrast, dusky, which means that darkness is relatively new to the human line. In religious art, Adam and Eve are always shown as light-skinned. Given the looks of today's Middle Easterners that was doubtful at best. The first modern humans, a hundred thousand years and more ago, were certainly black.

DNA hints that our new and swarthy appearance arose about a million years before the present. At just that time, our ancestors began to move from the forests to the sun-baked savannahs. Long legs and arms and a distinct nose (not found in chimps) also emerged, perhaps to cope with life in the sun. In addition, we lost our hair – perhaps to cool down – and dark skin was favoured as it protected against the harmful effects of ultraviolet light. The colour of the skin turns on the amount of a pigment called melanin.

The first hint about Eve's complexion came not from people but fish. The zebrafish is often used to study embryonic development. A certain mutant lacks the dark stripes that give the animal its name and is almost transparent. The gene responsible has been tracked down in both its mutated and its normal version. A search through our own DNA reveals an almost precise match; so close, indeed, that the human gene will reinstate a zebrafish's stripes when injected into a mutant embryo. The enzyme it makes shows a large shift in structure across the globe. A certain building block – an amino acid – is present at one point along the protein chain in 98 per cent of Africans, while in 99 per cent of Europeans it is replaced by a different version. The form found in Africa makes far more pigment than does the alternative. A large part of the shift in appearance between the inhabitants of the two continents hence emerges from a change in a single letter of the genome. The length of DNA involved varies not at all in its functional section throughout Africa, as a hint that dark skin was strongly favoured when it first arose and that any later changes have been removed by selection. Europeans are more diverse, with a variety of forms of the crucial protein that give rise to black, blonde or red hair and to dark or to almost translucent skin. Fossil DNA shows that Neanderthals had their own, different, mutation in that segment of the genome, so that they too were probably white.

In a twist to the tale, the light skins of China and Japan evolved in a different way. The gene that bleached the Europeans played no part, for the locals bear the African, rather than the European form. The people of the Far East paled in their own fashion, and evolution picked up changes in quite a different set of genes. A certain segment of DNA, when it goes wrong, causes albinism – a loss of skin pigment – in Europeans. The loss of melanin from Asian skin comes, in large part, from a mutation in a different section of that gene. Several other parts of the melanin factory differ in structure between Africa, Asia and Europe. Most have small but noticeable effects on colour, which is why the children of a marriage between an African and a European vary from dark to light and do not resemble either of their parents exactly.

The earliest modern Europeans and Asians of forty or fifty thousand years ago were almost certainly black. Even the French cave-painters at Lascaux may have had that complexion, for their images of the aurochs, the giant oxen, are reddish, while those of the men who hunt them are darker. The first Englishmen – those who followed the ice as it melted – reached these islands thirteen thousand years ago. They too may have retained their African colour when they set foot on their new nation's shores.

Why does it pay to be black in Benin but fair in Folkestone? Everything we know about melanin is positive, while fair skins seem at first sight to do more harm than good. Melanin protects against skin cancer – and fifty thousand people develop that in Britain each year. Two thousand die. Light skin burns easily. That may sound trivial, but sunburn makes it hard to sweat and easy to overheat, which brings dangers of its own. In addition, melanin reduces the destruction of vitamins in the blood as they are exposed to the harsh rays of the sun. Fair-skinned women who sunbathe have reduced levels of a vitamin called folic acid, and their newborn children pay the price, for a shortage of the stuff causes birth defects. Given the problems of pale skin, something powerful must have changed us on the journey from the azure firmament of the tropics to the gloom of British skies.

Another vitamin was to blame. Vitamin D helps build bones. It controls the levels of calcium and phosphorus in the blood for it helps the gut to absorb them and rescues quantities of each element that would otherwise be lost in the urine. Oily fish, eggs and mushrooms are rich in the stuff and many governments now add it to milk or flour to promote their citizens' health. Vitamin D can, in addition, be made in the skin through the action of ultraviolet light on a form of cholesterol.

To do the job, the light must get in and melanin keeps it out. Africans have to spend several hours a day in bright sunlight to make enough vitamin D to stay healthy, but northern Europeans who expose their arms, head and shoulders for fifteen minutes at a midsummer noon can make enough to meet their needs.

A shortage of the stuff puts children in danger of the soft-bone disease rickets, which leads to a curved spine or legs and can cause severe disability. Sufferers may also experience seizures and spasms – a side-effect of calcium shortage – which can end in heart failure. Nine out of ten infants in Victoria's smoky and starved cities showed signs of the illness and rickets is still the commonest non-infectious childhood disease in the world.

Most young black people in the United States have low levels of the crucial vitamin and the condition is, as a result, three times more common among black Americans than in their white fellow citizens. As a youth the athlete O. J. Simpson suffered from rickets and wore home-made leg braces. On this side of the Atlantic, my own generation was saved by free cod-liver oil, but the modern world is not so lucky. In Britain, soft bones are back. A third of Asian and Afro-Caribbean children are short of vitamin D (for the former the fact that they are not allowed to uncover themselves is in part to blame). Severe deficiency is nine times more frequent in that group than in Europeans and one in a hundred of their children show signs of illness. Girls do worse, which is bad news later on, for their pelvis narrows and they find it harder to give birth. There have even been cases of shortage in affluent white children allowed to play in the sunshine – but protected from the dangers of ultraviolet with sunscreen.

The magic substance also helps to hold diabetes, arthritis, muscular dystrophy and heart disease at bay and protects against the spread of certain cancers, with a higher rate of lung and bowel cancer in cloudy places. Any change in skin colour that helped to generate more of the vitamin must have been most helpful on mankind's journey into the gloom. Natural selection noticed the new mutations at once and in cloudy places fair skin soon took over.

Selection has lead to many other upheavals in human DNA. Many of them emerged from shifts in our habits as we moved from ape to early human, and to modern man. Migration, shifts in diet and the rise of towns and cities all led to genetic change.

For nine-tenths of our history as a species, most people saw fewer people in their lifetimes than an average westerner now does on his way to work. Agriculture led to a population explosion, and *Homo sapiens* is now ten thousand times more abundant than is any other mammal of his size. In a world of pathogens and parasites, abundance is an expensive luxury. Epidemics have often cut our species down to size. They need large populations to sustain themselves, and migrants to spread the infection. The Plague of Justinian, which began in Constantinople in AD 541, put paid to a quarter of the people of the Eastern Mediterranean. The Black Death spread along the Silk Road from China in the fourteenth century and returned again and again to the teeming and filthy cities of the west. Two out of three Europeans died. Sickness is potent fuel for selection and many genes respond to it.

One illness shows its power better than any other. A third of the world's population is exposed to malaria, half a billion are infected and the disease kills five people a minute. The real attack began about ten thousand years ago, when men moved into – and cut down – tropical forests at a time of warm, wet weather. That helped mosquitoes to breed and the parasite to spread.

In Kenyan families, poor conditions – a marshy spot, too much rain, too many children – explain some of the variation in individual risk of illness, but genetic differences are behind at least a third of the overall chance of ending up in hospital. Some variants have a large influence and are soon picked up by evolution while others are more subtle. The most important involve changes in the red blood pigment, haemoglobin. A quarter of a billion people bear at least a single copy of a mutated version of the molecule. The best known is sickle-cell, a simple change in the DNA alphabet. The haemoglobin of those with two copies forms long crystals in parts of the body low in oxygen. The red cells take up a crescent shape that restricts circulation and causes pain, heart disease and worse. Those with a single version of the altered message are healthy, with half the risk of fever if infected and a ten times lower chance of serious illness. A third of all Africans are in that situation and the gene is common in southern Europe, in the Middle East and in India. It has arisen on at least four different occasions. Other such changes give a lesser protection in countries such as Bangladesh, while deletions of long or short sections of DNA do the same in the Middle East and Oceania. Once again, those who carry two copies of a damaged gene pay a severe price while people with just one are protected.

Many other genetic changes have been pressed into service against that unpleasant illness. The parasite uses a certain red-cell enzyme to fuel its machinery. Hundreds of millions of people bear a defective version, but in return gain a defence. A certain form of the parasite cannot get into cells that lack a particular attachment

site. Almost all West Africans have this variant. Elsewhere, a change
in the shape of the red cell baffles the agent of infection, while the
high salt and iron levels in African blood also fend it off. Dozens of
sections of the DNA are implicated in the fight against malaria and
many, no doubt, remain to be discovered. Large or small, each has
been picked up by the selection, which, just as in the evolution of
pale skins in Europe and Asia, has cobbled together a response step
by step.

Natural selection is always poised to deal with enemies as they
arise. Wherever it works, it leaves evidence, often indirect, that it
has passed by. Some changes in DNA alter the structure of proteins
while others do not. The ratio between the two is a crude test of
its actions, for useful sections of the genome are more likely to
accumulate change under the influence of selection than are the
non-functional parts. On that criterion, our lineage has experienced rather less of its attentions than has that of the chimpanzee.

Another clue to the action of Darwin's agent comes from the
blocks of genetic variants packed close to each other along each
chromosome. As a favoured gene – a new anti-malaria mutation,
perhaps, or a change in skin colour – is picked up and becomes
more common, it will drag along sections of DNA on either side.
The stronger and more recent the selection, the longer the segment that accompanies it. In Africa, both the gene for black skin
colour and that for sickle-cell sit in the middle of great sections of
double helix that vary scarcely at all from person to person. That
pattern hints that in each case the new mutation was seized upon
at once and spread fast.

In time such uniform blocks of DNA are broken up by the
random reshuffling of genes that takes place when sperm and egg
are formed, but the process can take a long time. A length of
DNA that is identical from person to person within the generally
diverse genome is hence evidence that selection is, or has been, at
work. The human and chimp genomes each have thousands of

such segments. One gene in sixty among the chimps bears that Darwinian mark but only half as many in humans, as proof that we have coped with new challenges in a manner that our close relative cannot. Man's ability to modify the environment to suit his needs has weakened the hammer blows of nature. Anti-malaria drugs now do what could be achieved only by expensive mutations. Thousands of years ago, our skin responded fast to a shift in climate, with a genetic change; but most people, black or white, now protect themselves against the sun in quite a different way, with clothes.

The loss of our native nudity was an early hint of the evolutionary talent that made us unique – the ability to respond to a challenge not with bodies but with brains. Clothing allowed us to spread across the world, for with its help we take the tropics with us wherever we go.

Adam and Eve, in their sultry paradise, were unashamed, but after the first (and least original) of all sins they made aprons to hide their nether parts. When did they first put them on? Lice hint at when garments were invented. Chimps and gorillas have lots and spend many hours grooming as a result. When humans emerged on to the sunny savannahs they lost their hair. The lice had a hard time and evolved to live in the few patches of habitat left. We now have three kinds, the head and the body louse, plus the pubic louse. The body louse is the only one that hangs on to clothing. The pubic version is closest to the lice found on gorillas and may have joined us from there. DNA shows that the other two evolved from a chimpanzee parasite which began its intimate acquaintance with our own bodies six million years ago. The body and head forms, in contrast, separated more recently – perhaps no more than fifty thousand years before the present. That may mark the moment when we first donned our vestments and gave a resourceful louse a new place to live. Men, their parasites prove, dressed themselves as they took their first steps towards the icy north.

Since then we have learned to cope with external parasites with insecticides, with cold with central heating and with noxious foods with kitchens. Each talent is a product of the contents of the skull, which are − like Adam's underpants − unique. Darwin noted that 'There can be no doubt that the difference between the mind of the lowest man and that of the highest animal is immense', and he was right. To understand human evolution we need to know how and why our brain, the most human of organs, is so different from that of any other primate and why and how our behaviour is even more so.

The structure is three times as big, and the cortex, the thoughtful bit, five times larger than that of the chimpanzee and the modern skull is several times roomier than that of three million years ago. Chimps are born with a brain almost as big as that of an adult animal while babies, whose brains are already larger than that of a chimpanzee, continue to invest in grey matter until they are two. Genes active within the human cranium have multiplied themselves when compared with those in other primates and one such, which when it goes wrong leads to the birth of infants with tiny heads, has evolved particularly fast. The nerves within the human skull are more connected to each other, and their junctions more sophisticated, than are those of the chimp and the structure is also busier at the molecular level. Even so, much of the DNA most active in that part of the body has changed no more rapidly than that at work in liver, muscle or scrotum.

The brain is expensive, for by weight it uses about sixteen times more energy than does muscle. That represents a quarter of the entire budget of the body at rest and means that we expend twice as much effort on the intellect as do chimpanzees. How can we afford such a luxurious appendage? Humans eat no more than other primates of comparable size but have a richer diet, with more meat and fewer roots and leaves, than do our relatives. As a result we need smaller intestines to soak it up. We also invest less in muscle than other apes and the enzymes that burn food are more efficient than theirs. All this began, like black skin, a million and more years ago, when people moved from forests to savannahs, travelled in larger groups and became better hunters with a meatier diet. The way to man's brain was through his guts.

Even so, today's organ of thought is no bigger than that of the Neanderthals. Fossils of their newborns show that they were born with a brain as large as our own, which grew even faster during infancy, but those creatures acted far more like apes than we do. Something more than an extra dose of grey matter has made us what we are. To quote Darwin: 'of all the differences between man and the lower animals, the moral sense or conscience is by far the most important'. A glance at our relatives shows how right he was.

Chimpanzees are nastier than many people like to think. They kill monkeys and are pretty unpleasant to each other too. Their sex lives would shock Queen Victoria and their ethical universe, if they have such a thing, is far darker than our own. They live in groups, but the groups break and reform as their members quarrel. Terror makes the world go round. Set up a task in which two chimps need to pull a rope to get a tray of food. They will, but only if they are out of reach of each other. Otherwise, the dominant animal attacks its subordinate even if neither then gets anything. Anger and greed destroy the hope of reward. What makes humans different is a loss of fear, odd as that sounds in a world where that emotion seems to be everywhere. When anxiety goes, society can emerge.

Our social skills begin early. A group of two-year-olds asked to find a piece of food after they saw it moved to a new place or turned to a new position or put in a box with a beep were pretty good at each job – but no better than adult chimpanzees, for both babies and chimps succeeded at about two trials out of three. When it came to the need to learn from others the babies won hands down. They became far better at each problem when they saw someone else solve it, or when an adult pointed to or gazed at where the food was hidden or made noises that told them they were getting warm. Each response demands an insight into another's inner sentiments. We have a lot more of that talent than do our relatives. The chimps took no notice of those who tried to help.

Chimpanzees can learn, but do not teach: like all apes, they ape but do not educate. In some places, adults fish for insects with a stick or bash nuts with a stone, and the young emulate them. Even so, the grown-ups make no effort to show the infants how to do the job, do not change their ways when youngsters are around and never check to see how well they are doing. Birds, with their bird brains, can do what a chimp does, for a budgie will pull out the stopper of a bottle of food if it sees another do the job, and crows are even smarter.

Real education asks for more. A good pedagogue can teach almost any subject as long as he keeps a few pages ahead of his charges and they respond to his efforts. Teachers also have insight into the mental lives of their pupils, into who understands the lesson and who does not, and know how to encourage them without their becoming bored.

The chimp's negligence about the next generation is a reminder that the minds of our hairy relatives are not much like our own. A competent teacher needs to understand what his students are thinking – and chimps do not: they have no more than a rudi- mentary 'theory of mind', as psychologists put it. We have lots, and it helps those on both sides of the lectern. Teenagers might doubt the fact, but no ape could ever become a schoolmaster.

The best way of reading a mind is to chat to it. Thomas Love Peacock invented a character called Sir Oran Haut-Ton, who learns to play the French horn but not to speak (he is elected to Parliament, where his silence gives him an air of wisdom). *Homo sapiens* is the eloquent ape. Even deaf children left in groups babble with their hands. Speech is the scaffold upon which society is built. No other primate can speak and all attempts to persuade them to do so have failed (Noam Chomsky, the theoretician of language, noted that it was 'about as likely that an ape will prove to have a language ability as there is an island somewhere with flightless birds waiting for humans to teach them to fly').

The origin of language is a cause of endless dispute which, given that just one creature can speak, may never be resolved. Darwin thought that perhaps it began with imitation: that 'some unusually wise ape-like animal should have thought of imitating the growl of a beast of prey, so as to indicate to his fellow monkeys the nature of the expected danger'. *The Descent of Man* also suggests that it could have started with love songs, and that speech was in part a side-effect of sexual selection. Perhaps it was; or perhaps it grew instead from the simple fact that we are social animals. Apes groom each other because the constant pacification calms them all down and cuts down the conflict that is never far from the surface. Big groups demand too much scratching time but reassuring sounds can placate lots of individuals at once. The savage breast might first have been charmed in that way; possibly, indeed, with song – which could be why some stutterers can sing a sentence when they cannot say it.

However it began, language makes us what we are. The ability to speak is coded for on the left side of the brain and plenty of primates have a brain almost as lopsided as our own. Even so, chimp tongues fill their mouths while ours are dainty in comparison. The human tongue has retreated down the throat. The language of Shakespeare is a complex set of sounds made as the space above the larynx flexes and bends. The anatomical changes leave evidence in the shape of the skull. Neanderthals had chimp-like mouths and could do little more than grunt. The first skull capable of speech emerged no more than fifty thousand years ago – not long before the explosion of technology that led to the modern world.

One British child in twenty has some form of speech disorder. A certain rare inborn abnormality makes it impossible for those who inherit it to cope with grammar. Baby mice with the same damaged gene make fewer squeaks than usual when removed from their mothers, and people with a version impaired in a different way are at risk of schizophrenia; of, like Saint Joan, hearing voices that are not there. The normal version found in humans differs in two of its amino acids from that in all other primates. It is foolish to speak of a gene for language but if the transition from animal to human turned on speech it may have involved rather few molecular changes. The situation is confused by the discovery that Neanderthals have the human version of the gene, which must hence date back to our inarticulate joint ancestor.

Wherever they came from, words are the raw material of a new kind of genetics, in which information passes through mouths and ears as well as through eggs and sperm. It moved us on from our status as a rare East African ape to the most abundant of all mammals. Ideas, not genes, make us what we are. Our DNA is not very different from those of our kin, but what we do – or say – with it has formed our fate.

Even so, the famous 'indelible stamp' is without doubt imprinted into the human frame. Modern biology shows that chimpanzees are even more like us than Charles Darwin imagined – but in no more than the most literal way. The strengths and the limitations of his ideas in deciphering what makes us human have become ever clearer as knowledge advances. His theory is powerful indeed but enthusiasts need to be reminded where its power comes to an end.

In 1926, the Soviet government sent an expedition to Africa. It was directed by Ilya Ivanovich Ivanov, famous for his work on the hybridisation of horses and zebras by artificial insemination. The Politburo hoped to do the same with men and apes, for the experiment would be 'a decisive blow to religious teachings, and may be aptly used in our propaganda and in our struggle for the liberation of working people from the power of the Church'. In Guinea, Ivanov obtained sperm from an anonymous African and inseminated three chimps – but none became pregnant. He then planned to fertilise women with chimpanzee sperm, but was not allowed to do so. Back in Russia he set out to do the same with a male orang-utan and a woman who had written that 'With my private life in ruins, I don't see any sense in my further existence . . . But when I think that I could do a service for science, I feel enough courage to contact you. I beg you, don't refuse me . . . I ask you to accept me for the experiment.' The orang, alas, died before its moment of glory and Ivanov was arrested and exiled to Kazakhstan, where he, too, met a childless end.

Americans anxious to stop research in human genetics once attempted to patent the idea of a human–chimp hybrid in order to whip up protest. The application was denied on the equivocal grounds that the US constitution does not allow the ownership of human beings (whether the cross-breed would have that status was not discussed). Artificial fertilisation of chimpanzee egg with a man's sperm may now be feasible (although claims to have produced a 'humanzee' are fraudulent) but is universally seen as beyond the pale. The problem is not one of biology, but of what it means to be human. A hybrid between a chimp and *Homo sapiens* makes too ready an equation between our apish bodies and our immortal minds.

Charles Darwin was well aware of the limits of his own theory. As he points out in the famous last sentence of *The Descent of Man*, men and women possess noble qualities, sympathy for the debased, benevolence to the humblest and an intellect which penetrates the solar system. All that does not change the fact that in our bodily frames, most of all when reduced to chemical fragments, we bear the indestructible mark of our humble ancestry.

Some people despise his science as a result, because it appears to destroy man's special place in nature, but they fail to understand what evolution is all about. Biology, in its proof of our kinship with chimpanzees, underlines its irrelevance to ourselves. The double helix does not diminish *Homo sapiens* but sets him apart on a mental and moral peak of his own. The theory of evolution does not render us less human than we were before. Instead the insight it provides into man's place in nature has made us far more so than we ever realised. A century and a half after Queen Victoria's disagreeable visit to Jenny the orang-utan, I gave a talk at London Zoo which pointed this out – and most of the apes agreed.

CHAPTER II

THE GREEN TYRANNOSAURS

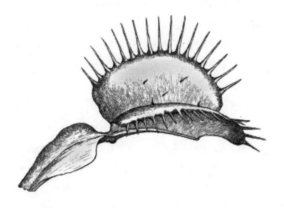

Soaring above southern Venezuela is a hidden landscape: the sandstone plateau of Mount Roraima, an inaccessible peak that is most of the time shrouded in mist. Arthur Conan Doyle used the place, or one very like it, as the location for his 1912 book *The Lost World*, a tale set in a land of evolutionary imagination, a place of dinosaurs, ape-men and primitive humans, ready to be explored by the irascible Professor Challenger. It was a fearsome spot but the bearded Englishman lambasted the lizards and saved the savages, as any Edwardian reader would expect.

Conan Doyle was born in the year of *The Origin*. By his fifty-third birthday, the theory of evolution had become so widely accepted that a literary hack could use it as the centrepiece of a work of fiction. Conan Doyle, who had read the reports of the British explorer who discovered the unique island in the sky, seized the chance and his book sold hundreds of thousands of copies to a well-primed public.

In reality the dinosaurs had gone from Roraima millions of years before and the local 'savages' never made it to the top. Even so, its remote summit is a genuine lost world, not of giant anthropophagous lizards or man-eating apes but of unobtrusive plants with the same dietary habits. Those green carnivores turn for food not to human flesh, but to insects. They must do so or starve.

Their habit is widespread. Almost six hundred insect-eating species, from all over the world, and from a wide variety of groups, have now been discovered. Their way of life has evolved on many occasions, and the tactics used to trap and digest prey are varied indeed. Separate lineages, from quite different places in the evolutionary tree, have taken up an identical diet and have come to the same solutions to find, digest and absorb their food. Charles Darwin had used such convergent evolution, as the process is known, as evidence for natural selection in *The Origin of Species*. The similarity of certain Australian marsupials to true mammals elsewhere in the world, or of wings in birds and bats, was, he saw, powerful proof of its action. Unrelated creatures faced with the same challenges adopt structures and habits that look similar but have different roots. As he pointed out, life can reach the same end through quite different pathways: 'in nearly the same way as two men have sometimes independently hit on the very same invention, so natural selection, working for the good of each being and taking advantage of analogous variations, has sometimes modified in very nearly the same manner two parts in two organic beings, which owe but little of their structure in common to inheritance from the same ancestor'.

Now we know many such examples – flight not just in birds and bats but in squids, fish, dinosaurs, flying squirrels and the marsupial sugar-glider of Australia (not to speak of the flying snake whose flattened body allows it to glide for many metres from a tall tree). We ourselves are not immune to convergence, for plenty of creatures have lost their hair, grown their brains, or even – as in the meerkats of Africa, who instruct their infants how to eat poisonous insects – stood upright and gained some simulacrum of the ability to educate.

Evolution in response to a common challenge has been so effective that certain creatures once assumed to be close relatives because they are so much alike in form are in fact not real kin: the vultures of the Old and New World, similar as they appear, do not have a recent common ancestor, for the former are eagles and the latter storks. Anteaters and aardvarks, lions and tigers, moles and mole-rats – all hide a bastard ancestry beneath their shared appearance. The process goes further. On Roraima itself, for unknown reasons, melanism is rife among unrelated organisms, and the rocks harbour black lizards, black frogs and black butterflies. The mutation responsible for black melanin pigment is the same, or almost so, in zebrafish, people, mice, bears, geese and Arctic skuas (and perhaps even in lizards and frogs), and has been picked up by natural selection in each. Within the cell, too, shared evolutionary pressures have produced enzymes with distinct histories that have settled on an almost identical DNA sequence in the active parts of the molecule. On a more intimate scale, the complicated chemical used as a sexual scent by certain species of butterfly also does the same job for elephants (which is riskier for one partner in the relationship than for the other). Evolution towards a common plan is just as rife among plants. The cactuses of the Americas – spiny, thick-skinned and globular – resemble the Euphorbias of South Africa, but have no more than a distant affinity to them.

Just after the publication of *The Origin*, Darwin began to work on a botanical lifestyle that, as he soon found, drags a great diversity of unrelated species into a shared set of habits. His interest began in 1860, when he visited Hartfield in Sussex, on the edge of Ashdown Forest, the home of his sister-in-law Sarah Elizabeth Wedgwood (and later the birthplace of Winnie the

Pooh). There he saw thousands of sundews – small clumped plants, with a sticky surface that traps insects. Some had as many as thirteen victims on a single leaf. Most of the prey consisted of small flies, but some victims were as large as a butterfly and he was told that the traps could even catch dragonflies. As so many sundews were present, the numbers of insects slaughtered must, he calculated, be prodigious. Each leaf had scores of glands held upright on fine hairs. They exuded shiny globules of liquid even on dry days and entangled any small creature foolish enough to land upon them. The sundew, he found, had feeble roots – evidence that most of its nutrition did indeed come from its gruesome way of life. He brought some specimens back to his greenhouse and began to explore how they did their job. It was the first step in a decade of work that produced a powerful vindication of his claim in *The Origin* that natural selection could, starting from different places, end up with much the same result.

In 1875, he published a book, *Insectivorous Plants*, on the subject. It deals not just with sundews but with a variety of such creatures from across the world, some from the area of Roraima itself. Darwin soon found many similarities among the various species that have taken up the habit. A closer look showed that many of their adaptations are also present in the other major kingdom of life. Emma noted in her diary when he was at work on a certain insectivore that 'I suppose he hopes to end in proving it to be an animal.' Her husband was so astonished by such parallels that he wrote to a friend that 'I am frightened and astounded at my results.' The aesthete John Ruskin said, in contrast, that 'with these obscene processes and prurient apparitions the gentle and happy scholar of flowers has nothing whatever to do. I am amazed and saddened, more than I can care to say, by finding how much that is abominable may be discovered by an ill-taught curiosity.' Darwin's curiosity, ill-taught or not, added another plank to his evolutionary edifice.

★

Roraima and the flat islands of rock around it are ancient indeed. Its sandstone peak – once part of a wide and barren plain, most of it now eroded away – is almost two billion years old. In the context of its immense history the terrible lizards went not long ago and the humans clustered around its base arrived in an evolutionary yesterday.

Its unique vegetation has been on its rain-soaked flanks for longer than either. The plants have seen the slow passage of time and have changed to match. A third of them evolved upon the mountain's lonely rocks and are found only there. Their native land is a hungry place. Constant downpours eat at the soil and strip what remains of nutriments, which are tipped down some of the highest waterfalls in the world. Worst of all, the rain washes away the nitrogen that every tree, shrub or flower needs to grow. Sandstone peaks, deserts, dunes, bogs, pine forests, Mediterranean scrublands and more – all are short of that element and each, distinct as it looks, and different as its inhabitants might be, has evolved a set of inhabitants whose battle for existence is focused, in a variety of ways, on the need to find it. The struggle for nitrogen shows – even better than the multitude of ways in which life has taken to the skies – how natural selection can reach the same end with different means in creatures from quite separate parts of the biological universe. Plants, animals, bacteria and fungi are all drawn together in their shared hunger for the element and all have become entangled with each other in the struggle to find it.

The gas itself makes up four-fifths of the air but plants cannot extract it directly. Their growth is often limited by a shortage of the stuff. Many manage by soaking it up from the soil. They forage like hungry animals with their roots, which stretch further and further as the essential item runs short. Below the surface, much of the element is bound into compounds that refuse to give it up.

However adept they may be as foragers, many plants live on

soils with so little nitrogen that they cannot survive without help. They are forced into contracts with other creatures that donate the vital element. The natives of Roraima eat insects and soak up the nitrogen in their flesh. The trade is one-way, for the plants kill the animals. Sometimes, in contrast, the treaty seems positive, for both parties appear to gain. In fact, even the several apparently amicable associations in the nitrogen market are based on greed and expedience. The fight for the crucial chemical is fierce and has led to shared adaptations that straddle the whole of life.

All animals fall prey to the vegetable world in the end as their dust returns to dust, but in the starved landscape of the Lost World the vegetation cuts out the middleman and devours the local wildlife directly. Natural selection has tinkered with leaves, roots and other parts to come up with the equivalents of teeth, gullets, stomachs and intestines, to draw the machinery of the botanical world close to that of animals.

Charles Darwin's book on the insect-eaters sold less well than had Conan Doyle's fictional lizards but such creatures behave in a way beyond the imagination of the most fanciful novelist. *Insectivorous Plants* raised biological questions that resonate beyond the universe of the carnivores. At first Darwin doubted the value of his own experiments and wrote to a colleague that 'I must consult you some time whether my "twaddle" is worth communicating', but he soon became an enthusiast. His book is a masterful narrative of the ingenuity of existence.

Plants that eat flesh attracted curiosity and hyperbole long before Darwin's day. They still do, even if the Australian version that feeds on rabbits has yet to be confirmed by science. The sundews of Ashdown Forest have a natural flypaper that holds its victims with a syrupy glue. The leaves curl round to entangle them before they meet a sticky end in its sinister digestive globules. Their secretions give the plant its name; as Henry Lyte wrote in 1578 in his *Nievve Herbal*: 'This herbe is of a very strange nature and marvelous: for although that the Sonne do shine hoate, and a long time thereon, yet you shall finde it always moyst and bedewed.' *Insectivorous Plants* also includes experiments on the Venus flytrap, which modified some of its leaves into a 'horrid prison', and on many other kinds forwarded to Down House from afar.

The insectivore habit has evolved in around a dozen distinct lineages, and Darwin saw most of them. They represent but a small fraction of the quarter of a million kinds of plants with flowers, but their habits, and their origins, are varied indeed. Some of the flesh-eaters are relatives as close as the anteater and the pangolin (long-nosed predators of ants from the Americas and the Old World respectively, the former kin to armadillos and the latter to dogs and cats) but others are as different as an anteater is from an insect-eating lizard or a bird such as the swallow.

The sundew belongs to a group of a hundred or so species whose centre of diversity is in Australia. It has a relative called the butterwort that hunts in the same way. The flypaper habit has evolved on at least five independent occasions, to produce the Australian 'rainbow plants', so called because of the sinister sheen of their leaves, and many more. Some kinds are three metres high and some tiny, and they are found across much of the globe, from Alaska to New Zealand. Europe has just three of the hundreds of species known. DNA shows that even the sundew and a very similar species from Portugal have evolved insectivory of their own accord, as each can trace affinity to closer relatives that do not indulge in the pastime.

A second trick remarked on by Darwin is to swing a door shut upon the prey. The most familiar jailer is the Venus flytrap, brought to attention in 1768 as the 'fly trap sensitive' by Arthur Dobbs, Governor of North Carolina, who sent the first specimen back to Britain (and was in addition the first person to record the movement of pollen by bees). The botanist William Bartram saw the 'ludicrous' plant on his travels through the Carolinas: 'This wonderful plant seems to be distinguished in the creation, by the

Author of nature, with faculties eminently superior to every other vegetable production . . . We see here, in this plant, motion and volition.' The 'irritable principle in vegetables' was much commented on even if some denied that any plant would lower itself to prey on an animal. Linnaeus, the great classifier, insisted that the flytrap always let its prisoners go, while others claimed that the insects trapped within were sheltering from frogs. It even became part of legal philosophy. The social theorist Cesare Lombroso, who believed that crime was a biological throwback beyond the control of those responsible, felt that flytraps marked 'the dawn of criminality'. They 'establish that premeditation, ambush, killing for greed, and, to a certain extent, decision-making (refusing to kill insects that are too small) are derived completely from histology or the microstructure of organic tissue – and not from an alleged will.'

The flytrap – and just a single species is known – lives in nitrogen-poor swamps in the Carolinas, its native home. It was given the erroneous Latin tag of *Dionaea muscipula* (which means 'mouse-eater' rather than 'fly-eater' as intended). Its popular name among the colonists was 'tipitiwitchet', then a slang term for female genitalia, because of its supposed resemblance to that organ, (although to avoid vulgarity Thomas Jefferson used the label 'Aphrodite's mousetrap' when he added his specimen to his collection). Each bears up to a dozen traps, each made of a much-modified leaf. Darwin himself considered the creature to be 'one of the most wonderful in the world'. Its trap closed on its prey with spines that interlocked like the teeth of a rat-trap, rather than gluing them to its leaves, but its sensitivity reminded him of the sundew's quite different strategy.

Just two snap-traps are known: the Venus flytrap itself, and the so-called waterwheel plant (again just a single species, but found scattered across the world) which does the same under water, on a smaller scale.

Other freshwater flesh-eaters use another method: a lobster pot, a snare with a one-way entrance valve. A separate group, found both on land and in fresh water, has tiny capsules or bladders that, when touched, suck in prey with irresistible force.

The bladderworts, as they are called, have hundreds of species, found everywhere except in Antarctica. They prime their trap by pumping water out across its wall. Another kind found on moist rocks in South America and parts of Africa specialises in single-celled animals, protozoa that swim into tiny slits in its specialised underground leaves. It shows a microscopic kind of carnivory and Darwin had speculated that it was indeed a meat-eater, although he had no idea of its food.

The insect-killers of Roraima use a different trick, a pitfall based on rolled leaves with margins sealed together. The fatal ambush is covered with a slippery glaze and decorated with a nectar bait. The sixteen species known from that peak are relatives of heathers and their mountain allies. Some are a metre tall, some tiny. Other pitcher plants of the New and Old Worlds and of Australia also use rolled-up leaves to snare their prey and some are big enough to take mice. They come from quite a different section of the botanical kingdom. Their capital is in the New World, which has more of those baleful creatures than anywhere else. The cobra lily of the western United States has a mouth said to resemble that of a snake. It feasts on ants. Others among its kin have a flared cover that shelters the jaws of the trap and keeps water out. A different group with the same general appearance, the monkey cups from around the Indian Ocean, make their pitfalls as structures that spring from a leaf's mid-rib and are held at the end of a long tendril. They get their taxonomic name, *Nepenthes*, from the restorative drug given to Helen of Troy. Linnaeus was impressed by them: 'What botanist would not be filled with admiration if, after a long journey, he should find this wonderful plant. In his astonishment, past ills would be forgotten when beholding this admirable work of the Creator!' Yet another pitcher is found in Western Australia. Its traps look rather like old shoes and the plant is unrelated to any of the other pitfall-makers.

DNA reveals some unexpected affinities among the pitchers, for the Old World kinds are in fact more related to sundews and Venus flytraps than they are to their New World equivalents. In addition they are quite close to a group of non-carnivorous lianas of tropical forests, and find more relatives in a larger class that includes rhubarbs, spinach and beet.

An even more distinct group, the bromeliads – relatives of the pineapple and in quite a different subdivision of the kingdom from the other green carnivores – also make leafy containers that fill with water. They live in the tropical forests of the Americas. There may be more than a hundred thousand in every hectare, most of them stuck to trees. The vessels made by their fused leaves generate a huge series of tiny lakes, in which a variety of creatures find a home. At first the bromeliads appear benign, for they lack the digestive enzymes found in the other pitfalls. Tadpoles, insect larvae, twenty-five-millimetre-long salamanders and tiny crabs all live in the liquid within. In truth, they have a darker side. Each watery island is full of conflict, and their proprietors gain nitrogen from the corpses of the creatures that are killed there and are broken down by bacteria. One kind has already taken a step to true carnivory with digestive enzymes of its own.

Other meat-eaters, unknown in the nineteenth century, are bizarre indeed. Certain soil fungi devour nematodes, worm-like creatures far larger than themselves, with a lasso that snaps shut within a tenth of a second, strangles the animal and gives the predator a rich source of food. Others do the same job with a sticky pad, while the familiar and tasty inkcap mushroom puts out spiked and lethal balls that puncture its prey and allow the fungal spores to grow within it. A hundred and fifty fungi that snack on flesh are known and there is a whole world of hunting mushrooms ready to be discovered.

★

Wherever they sit in the botanical world, a hard life in a hungry place has pushed every flesh-eater towards a similar set of expedients. Like cactuses they succeed where others fail, in their case because of a shortage of food rather than of water. The cost of success is specialisation. Their habit can be expensive and their way of life fragile, for traps cost a lot and force a reduction in the investment in roots or leaves. Insectivores often find it hard to cohabit with other species, which means that vast areas are carpeted by them alone. In addition, they are forced to seek open sunlight, for their leaves are too feeble to cope with shade, and do best in places where fires sweep through and wipe out the opposition.

The insect-eating package is sometimes lost when conditions change. Some are committed to it. The Venus flytrap is itself trapped into its narrow way of life, for it gains around three-quarters of its nitrogen from insects. The cobra lily is not far behind, and the snares of both those species are elaborate and hard to build. Other plants are more adaptable in their ways. Almost all the carnivores have some chlorophyll, the stuff that makes leaves green, but often no more than half that found in normal species. They gain some energy from the sun, albeit at reduced efficiency. The sundew and many of its fellows have reduced roots as not much food is available in their native swamps, but they can soak up a little. Radioactively labelled nitrogen shows that up to half of the nitrogen taken up by a typical individual comes from soil rather than from flesh.

For most insectivores, the prey are more important in the summer when they are abundant – and a large part of what they provide goes to make flowers, expensive as they are. The bladderwort makes traps only at the height of that season, when insects are common and the time has come for sex. The pitchers of New England make more traps in the bogs with least nitrogen, but put effort into ordinary leaves when the water contains more of that mineral. Pitchers make two kinds of leaves, either modified to make a trap or large and flat to soak up sunshine. When nitrogen is added, the plants make more of the latter, for a shortage of the element no longer limits their growth. Others, too, play the mineral market. The sundew produces less slime than normal when given a decent dose of fertiliser. All this suggests that carnivory is a luxury that is abandoned whenever a cheaper source of nitrogen becomes available. That has happened many times, for DNA shows that several vegetarians have evolved from carnivore ancestors. Certain pitchers from Borneo now soak up nutrition from dead leaves or from bird excrement that falls into their flasks instead of from insects.

The first botanical carnivores evolved long ago. A famous fossil bed at Yixian, in north-east China, from around a hundred and twenty-five million years before the present, yields dinosaurs with feathers – together with a small pitcher quite similar to those of today, with lures to attract its prey and glands ready to soak up their remains. The bed dates from a time close to that of the origin of flowers themselves. A tree of relatedness of that entire group based on DNA suggests that perhaps the sticky traps evolved first, while pitfalls and snap-traps came later – which, if true, pushes the habit even further into the past. Even the nematode-eating fungi have left a hundred-million-year-old fossil, trapped in amber in a French quarry.

Our insight into that motley set of unrelated creatures was transformed by Charles Darwin, who raised – and answered – biological questions that resonate far beyond their own narrow universe. He began a systematic survey of how the various members caught their prey, digested it and absorbed its goodness. The results, he wrote, were 'highly remarkable'.

First, he found that sundews grew far better when fed with insects than when starved, although they could survive for a time on a vegetarian diet. Their sensitivity was impressive. Even a tiny gnat with its 'excessively delicate feet' was enough to set off a reaction. A gland moved within ten seconds of the arrival of a meal. The impulse to do so soon spread through a whole leaf. Within an hour, a mass of tentacles began to bend towards the prey. The trap moved faster on warm days but light and dark made no difference. The tentacles had no sense of smell for an object had to touch the surface to induce the effect, but they could taste, for they held on to pieces of meat for longer than to bits of glass, cork or hair. Water, tea and sherry did not excite them and neither did a firm prod with a twig, but even a minute particle of living material led to some response.

What woke the sundew up was nitrogen. Urine (presumably the great naturalist's own) did the job, while tea did not. Ammonium carbonate − *sal volatile*, its ammonia used to stimulate fainting Victorian ladies − contains that essential nutriment and elicited a response almost at once, even when diluted. Just a twenty-millionth of a grain of nitrogen salts in solution − less than a thousandth of a milligram − was enough to stir the sundew's interest.

The Venus flytrap solved the hunter's problem with quite a different set of tricks. Its prison doors slam shut within a tenth of a second of being touched − which means that it makes among the fastest of all plant movements. As soon as the prey arrives it disturbs the sensitive hairs that cover the trap. A wave of electrical activity passes across the surface at around ten centimetres a second. The wall of each modified leaf, with its two hinged parts, has cells kept full of liquid at high pressure and the flytrap uses its feeble powers to pump them up, with many hours of effort needed to reset the snare each time it has been used. The trap wall is curved outwards when open, and inwards when shut, with a fine balance between the two stable states. A squeeze with the fingers causes it to snap shut in the same way as a pea pod pops open. The elastic energy stored in the curved shell of an open trap is released in a sudden rush and it slams closed.

The flytrap's trigger − which evolved down quite a different path from that of the sundew − gives further hints about how information from outside is translated into action. Darwin found

that it was as alert to a sudden touch as was the gland of its sticky
fellow carnivore, but responded less to prolonged pressure. Two or
three quick taps within thirty seconds of each other, rather than
just one, were needed to spring it, perhaps to avoid disturbance by
wind-blown grains of dust. Rain had no effect.

The pitcher plants had evolved another ingenious way to catch
their prey. Many befuddle the insects with signals that draw them
to a promise of food or sex but in fact deliver death. Certain
pitchers have a nectar-laden 'spoon' near the mouth while others
generate motifs that are irresistible to bees and other insects, such
as dark centres with stripes that radiate out and look rather like
flowers. Some have backward-pointing hairs within the pitfall to
prevent escape. The cobra lily has a guileful way with its victims,
for its walls are decorated with clear patches that persuade the flies
to stay and beat against a window until they drown rather than
flying upwards to gain freedom. Once they have fallen into the
liquid their fate is sealed for it contains a syrup from which escape
is impossible.

The slippery wax that covers the inner surface of many pitfalls
and causes insects to plunge to their doom has revealed its secrets.
It has two layers. The lower section is made of stiff foam, while
the upper consists of a sheet of loosely attached and brittle plates.
These break off and clog the prey's hooked feet, which then skid
on the foam below. Other species keep their surface moist with a
series of fine ridges that trap water or nectar and cause any insect
that lands to aquaplane into the void below. As a result, certain
kinds can catch their quarry only in rainy weather. At other times
an insect can sit happily on the rim. That, perhaps, lulls it into a
false sense of confidence when it comes for food until, one day, it
falls to its doom.

To trap a fly is just the first step to sucking out its goodness.
The animal must be digested and its nitrogen absorbed. Once
again, the insect-eaters use different means to achieve a common
end. As they do, they seem, once again, to approach the habits of

animals. A sundew leaf with a series of tentacles all pointing towards an item of food led Charles Darwin to 'imagine that we were looking at a lowly organised animal seizing prey with its arms'. Soon the whole leaf closed up to make a 'temporary stomach' that smothered the prey as each hair pumped out digestive droplets, the 'dew' that gave the plant its name. By a 'curious sort of rolling movement', rather like that of the human intestine, the unfortunate victim was propelled towards the centre, where more hairs awaited. As it expired, the secretion changed from a mere glue to an acid liquid. The sticky fluid could, he found, digest living material of many kinds.

The magic liquid was pumped out from the base of each hair. There were marked changes in the internal structure of each cell as they prepared to make the digestive juice, with the accumulation of masses of purple matter after a dose of meat. The sundews were a first hint that cells can communicate with each other, for a colour change could be tracked as the message spread across the leaf. The agitation spread just as in animal nerves, although nerves act faster and show no visible changes as they do their work.

What path did the information take as it travelled through the leaf? Various vessels traverse it, but tentacles close to and distant from such channels behaved in the same way. The best that Darwin could come up with was that the motor impulse involves the passage of chemicals of some kind but what these were he had no idea.

In those days all that was known about nerve transmission was that it involved what he called an 'invisible molecular change that is sent from one end of the nerve to the other', but there was no evidence of what that might be. His experiments with poisons on sundews provided a hint of what later became a central truth of cell biology. Some, from arsenic to strychnine, were as pernicious to plants as to ourselves, while others, cobra venom included, were not. Morphine and alcohol, with their noticeable actions on the human nervous system, left the sundews indifferent – but salts of potassium and of sodium had opposed effects, for the former caused the leaf to move while the latter was fatal. That observation presaged the later discovery that a balance between the two elements on either side of the cell membrane is behind the electrical activity of both plant and animal cells, nerves included.

The flytrap with its need for a double tap before closing must, in some sense, remember the first before it responds to the second. It was suggested to Darwin by Burdon Sanderson, Professor of Physiology at University College London, that a chemical or electrical mechanism was involved. He was the first to find that – just as in animal muscles when they contract – the voltage altered when it snapped shut. A complicated long-chained sugar causes the prison walls to close when applied in minute quantities to the leaf. It builds up at speed after the trap is touched, but is broken down almost as fast. Only if the second tap arrives in time does it reach a concentration high enough to trigger off a response. The memory molecule – which is what it is – activates channels in the cell membrane that transmit sodium and potassium ions. As it does, it generates an electrical signal that fires off the poised cells. As in nerves and muscles, the movement of calcium ions is also involved. The ambush can also be induced to snap shut with human chemical messengers such as adrenaline and certain other nerve-transmitters, which are themselves molecules that work by reaching a threshold.

Concerned as ever with the state of his own intestines, Darwin turned again for advice about the sundew's digestive juice to Burdon Sanderson. Its power to break down protein had long been known. The sticky secretion of butterwort was once used to make 'ropy milk', a sort of yoghurt in which the milk was curdled under the influence of its digestive enzymes. Herbalists still insist that the substance is useful against tuberculosis, asthma, intestinal pain and the chapped udders of cows (in the Netherlands it was once popular as a hair pomade).

The two scientists established that the exudate contained a

series of organic acids related to vinegar, together with an enzyme. When both were present – but not when just acid or enzyme alone was available – insect flesh was broken down. The sundew stomach, if such it could be called, hence showed close parallels in its actions to our own, which itself contains both acids and enzymes. That, Charles Darwin felt, was a 'new and wonderful fact in physiology', for it brought the plant and animal kingdoms together.

The digestive enzymes of the insect-eaters have now revealed more of their secrets. Our own gastric talents are limited in comparison with theirs, for the plants cope with a diet that would give us all dyspepsia. Vincent Holt's neglected 1885 work *Why Not Eat Insects?* contained recipes for nutritious dishes such as *larves des guêpes frites au rayon* (wasp grubs fried in their nests). Some people do indeed eat larvae – in the Far East silkworms are popular – but some of Holt's suggestions, such as *phalènes à l'hottentot* (moths in butter) and *cerfs-volants à la gru gru* (stag beetles on toast), would be indigestible indeed. The Vietnamese who feast on water-beetles or scorpions have to throw away the tough outer shells as impossible to manage. Their botanical relatives can afford to be less fastidious, for their enzymes can break down all the prey has to offer, the hard coat included.

The insectivores use, as do our own intestines, a cocktail of chemicals, each of which digests a particular foodstuff. Sundews have an enzyme that cuts up nucleic acids, while pitchers – which can hold three litres of digestive fluid – have half a dozen distinct kinds that attack proteins, nucleic acids and other substances, together with a special protein that breaks down the insect skeleton. The Venus flytrap has equivalents of its own (and if a leaf overeats it may die from indigestion), as do almost all the other botanical carnivores.

For plants and people alike digestion is followed by absorption. The insectivores have refined their abilities to soak up a meaty soup, but at a price, for the leaves of conventional plants are, in the main, quite impermeable because of the need to conserve water. A creature that feeds through its leaves cannot safeguard itself in that way. The sundew has large pores in a generally waterproof skin that allow its insect broth into the digestive cells. In pitchers, the whole interior surface is thin or is scattered with holes that allow water to pass. The carnivores then face a dilemma, for as they suck in the liquid remains of their feasts across a porous leaf surface they are at risk of fluid loss from the same place. As a result, many are restricted to wet places.

Such plants have to make many other compromises. First, they face a conflict between sex and starvation. They eat insects but are also pollinated by them. To reduce the chance of error, flowers and traps open at different times, or on different parts of their parent's anatomy, or attract a separate set of visitors. Even so, the carnivores often devour their winged Cupids by mistake or, perhaps, because they are more valuable as a source of nitrogen than as a sexual aid.

What struck Darwin most of all about his insect-eaters – and he did experiments on the underwater kinds as well as those on land – was that all of them built their specialised machinery by picking up and using talents already found in species with more orthodox habits. They were a wonderful example of how evolution could make do and mend. Natural selection often scavenged its raw material from whatever was available rather than being forced to wait for what it needed to emerge anew.

Varied as the insect-eaters are, and distinct as their traps and their means of digestion and absorption might be, carnivory has always been cobbled together from pre-existing structures. All the species studied at Down House, and the many more now known, have modified the banal talents of their ancestors to reach their present state. All roots make mucus, and sundews themselves are related to tamarisks and knotweeds, which make lots of the stuff to get rid of salt or to fight off insects. Most plants are attacked by insects and some have evolved defences such as glue-covered hairs, or spines that can, when needed, be used for aggressive ends. The familiar blue Plumbago is notorious for its sticky flowers that trap its enemies and save it from attack. Many flowers close upon a pollinator before they release it and – as a hint of how the flytrap evolved its remarkable talent – plenty of plants can move their leaves or seed-pods, some at speed. The pitchers had less to do to develop a trap, for leaves fuse for many reasons and a variety of mutations in crops such as maize cause once-independent leaves to bond together.

Digestion, too, has its echoes in the more innocent parts of the botanical world. Some wild geraniums are covered in a secretion that can break down and absorb proteins placed upon their leaves. They are 'proto-carnivores', poised on the edge of that habit but not yet committed to it. Other parts of the digestive tool-kit are also lying about, ready to be used. Seeds and many leaves secrete enzymes to protect themselves against attack. Those of certain insectivores resemble others that, in most species, are found only within the cell, proof that the leap towards carnivory did not involve some novel chemical but just a talent to pump one out. The sundew enzyme that cuts up nucleic acids (a material abundant in insect cells) looks rather like those secreted by all plants after damage. That, too, was hijacked. In the same way, the enzyme used to chew up the hard outer husk of an insect is close to those made by other plants when placed under stress.

Most leaves can absorb some molecules, small and sometimes large, through their surfaces. Darwin himself found that even species that never eat insects, such as Primulas, could transfer nitrogen-rich nutriments like ammonium carbonate across their leaves. His observation led in time to the idea of 'foliar feeding'. Instead of adding fertiliser to soil, where it can be washed away or rendered useless through chemical change, the hope was to spray it on to leaves, whence it would enter. In alkaline places there can be plenty of iron and manganese – both needed for healthy growth – but they are bound into soil compounds that will not release them. For some organic gardeners the idea has become almost a cult, but it gives real benefit only when rare nutriments are in short supply. Some places lack zinc, or copper, or boron, all needed in minute amounts. A quick spray does a lot to help. The technique is of no use for nitrogen, which is needed in large quantities, for leaves are too small to soak enough up.

Carnivory is just a further step – a case of the biter bit – in the endless battle between the insects and their vegetable prey. Each of their tactics is in use somewhere else, for a different reason. All that evolution had to do was to put the package together. The end result can be achieved in quite different ways, each of which works reasonably well.

The relationship between a carnivore and its prey shows a clear divergence of interest. Even so, their conflict can sometimes shade into what looks like cooperation. Many insect-eaters depend on third parties to help them. The North American pitfall known as the Virgin Mary's Socks (from its purple colour and the footwear of the Pope) has no digestive enzymes of its own and depends on bacteria to do the job. A South African species that, at first sight, looks like a typical carnivore has sticky hairs that trap insects – but it gains goodness at second hand. A bug makes its home upon the leaves and feeds upon the corpses, and its excreta feed its host.

The struggle for existence between fly-trapper and fly is easy to observe. It is a microcosm of nature red in leaf and glue. Some insects, in contrast, live not as prey for plants, but – like the South African bug – in apparent harmony with them. Certain ants, too, defend their hosts against attack – a talent known to the fourth-century Chinese, who put their nests into lemon trees. As is true for the insect-eaters, the tie between the two kingdoms has prompted the evolution of some remarkable organs, each of which has emerged, like a snap-trap or a flypaper, from a distinct part of the plant's anatomy. A celebrated passage from *The Origin* reads: 'If it could be proved that any part of the structure of any one species had been formed for the exclusive good of another species, it would annihilate my theory, for such could not have been produced through natural selection.' The helpful ants might appear to be such a threat, but they support the idea of evolution and do not annihilate it. On the way they hint at how insectivory began, for they pay a good part of their rent not with aggression, but with nitrogen. They add a whole group of new members to the grand botanical convergence that faces the fertiliser problem.

Thomas Belt was an engineer and naturalist who spent five years in charge of a gold mine in Central America. Darwin called Belt's 1874 book *A Naturalist in Nicaragua* 'the best of all natural history journals'. It notes how some trees in the Acacia family have fallen into an association with certain ants, who protect them against grazing insects and mammals. As Belt saw, the balance of advantage is delicate indeed.

The total mass of ants in a patch of Amazon jungle is four times that of all its mammals, birds, reptiles and amphibians put together. Certain plants, there and elsewhere, have put them to work. Many tropical trees have hollow thorns which shelter the vicious insects, together with small structures filled with sweet and sticky material. Both parties benefit, for any creature that dares to browse on the tree is attacked and the ant gets a free meal. If its garrison is killed off with insecticides the tree is attacked by grazers at ten times the previous rate. The helpers also prune back branches of nearby trees that shade their host, and clean up the ground around its trunk, reducing competition for food. Some ants even poison nearby plants as they inject formic acid into the leaves. That then allows their own host to flourish on patches of cleared ground known to the locals as 'devil's gardens' and thought to be cultivated by an evil and cloven-hoofed spirit. In return the insects feed on secretions from the acacia's leaves and feed their young from what Darwin called its 'wonderful food bodies'. They also gain protection by laying eggs inside the hollow thorns.

More than a hundred distinct groups of tropical trees, and forty families of ants, have entered into such an alliance. The habit has evolved many times and – like insectivory – has enabled natural selection to pick up a diversity of parts for use in a novel way. The shelters are based on thorns, on hollow stems or on rolled-up leaves, or on special pouches made on the surface of the leaf. Once again, evolution makes do and mends, as it must.

The details of the liaison give proof of Darwin's insistence that natural selection allows nobody a free lunch. At first sight, the bond between ant and trees is based on a shared dedication to a common end. In fact, each tries to get the most out of the arrangement while putting the least possible in. Their tactics hint at how the tie between the botanical carnivores and their prey may have begun.

Often, the special food is produced only when enough ants are around to make it worthwhile. Some trees are even more parsimonious. The whistling thorn of Kenya, which gets its name because the wind howls through its hollow thorns, uses ants to keep hungry giraffes at bay. It gives shelter, but no food, to its resident army. Ants cheat just as much, for some get a meal from the honeydew made by scale insects that feed on sap (and do no good to the tree) while others have little interest in attacking herbivores. Some species are even more selfish, for they castrate their host by chewing off flower buds to ensure that it does not waste its efforts on show, but puts out new and tasty shoots, with their free food instead. The tree fights back with a chemical that keeps the ants off the flowers. Yet another insect destroys its host's food bodies to dissuade more aggressive ants that might protect the tree but throw off the resident.

The ants gain sugars, based on carbon, from their host – but their corpses and those of their prey provide precious nitrogen to the tree. The shelters have thin walls through which the excretions of the residents, or the remains of their bodies, are picked up by the host tissue. Some acacias take, as a result, nine-tenths of their nitrogen from their insect visitors. It would not be too hard to transform an arboreal ants' nest into a trap that soaks up nitrogen while giving nothing back.

The acacias, like the sundews, are nitrogen hunters that depend on other creatures for help and, with the entry of the ants, make that hungry clan – already diverse – even wider than before. A further look around the botanical world shows that the tactics of acacias or Venus flytraps are feeble when compared with the ingenuity shown by other species. Many plants thrive in what would otherwise be famine conditions thanks to a series of obscure but intimate associations with other creatures in the search for the essential element. They negotiate not with insects – which, as animals, are quite close to plants in evolutionary terms – but with bacteria and fungi around their roots that pull the gas from the air and receive food and shelter in return. That habit is central to the survival of life on earth. It represents a series of evolutionary convergences between minute creatures far less closely related to each other and to their hosts than are insects.

The roots of many plants secrete chemicals that attract bacteria able to transform bound nitrogen into a more digestible product. They then soak up their invaluable wastes. Peas, beans and certain trees have entered into a closer arrangement with specialised 'nitrogen-fixing' bacteria that combine nitrogen gas in the air with hydrogen to form ammonia and other compounds which can be soaked up by roots. Many of the insect-eaters and ant-exploiters, with their spectacular adaptations above the ground, also depend on a similar pact with tiny aliens within their roots.

Farmers take advantage of such arrangements when they rotate their cereal crops with legumes such as clover and soybeans, all of which have close relations with nitrogen-fixers. Together, such plants now generate half the nitrogen used on the world's farms. Without them we would starve. Their bacterial allies make a special enzyme which forces the sullen molecules of the gas into a marriage with the active hydrogen ions made as food is broken down. The reaction consumes a great deal of energy and costs both the bacteria and its host a lot.

Before today's technical developments in biology, the bacterium involved, and the protein that does the job, looked more or less the same in each of the thousands of species that indulge in the habit. They are not. Just as in the insect-eaters, many unrelated plants, and even more of their minute helpers, have taken up the pastime. DNA shows that bacteria themselves, unimpressive as they might appear, are more diverse as a group than are the two kingdoms of animals and plants put together. The nitrogen-fixers span a good part of the spectrum of bacterial life. They are joined in their helpful habits by fungi, who are themselves more related to ourselves than are bacteria, and by members of a quite distinct group of single-celled beings known as the Archaea that teem in hot springs, deep sea vents and the soil. The sea, too, is itself full of a great variety of gas-fixers, most of them little understood.

Most of the bacteria involved live for most of the time alone. When they come into contact with a root of the right kind, a certain sugar locks into a receptor on its surface. The nitrogen-fixer squeezes its way in and its host's cells divide to produce a nodule filled with descendants of the invader – a billion or more from a single founder cell. Both parties benefit for the plant provides fuel for the hard chemical work needed to drag the crucial element from the air while the bacteria churn it out in a form that can be used by the other member of the consortium.

The association between the two emerged, in evolutionary terms, not long ago; just after the destruction of the dinosaurs. There was, at about that time, a sudden outburst of carbon dioxide and a spike in temperature, both of which favour plant growth – and meant that a sudden shortage of nitrogen made it worthwhile to enter into the arrangement. It has evolved again and again in distantly related families. Alder trees (but not their close relatives the beeches) have root nodules that contain bacteria better known as the producers of the antibiotic actinomycin. With their help the trees grow on starved soils such as those on dunes or mountains. Tropical ironwoods have the same association as do a few members of the rose and pumpkin families. Liverworts, certain ferns and the giant rhubarb of Brazil all benefit from the ability to use single-celled creatures to soak up the vital gas.

The plants that have come up with that solution are diverse indeed. The creatures that do the work are far more so. The nitrogen-fixers within the roots of beans, clover and their relatives have been widely studied because of their economic importance. Hundreds of different helpers have been pressed into service. Some are tied to a single host – or even to a particular cultivated variety of peas or beans – while others are promiscuous. Under a mask of similarity, the biochemical mechanisms involved, the molecules that signal willingness to enter into an association, the amount of food provided and the rewards paid are diverse indeed.

Like ants on acacias or insects that buzz around a pitcher, the system shows a fine balance between cooperation and conflict. Some bacteria enter their hosts through wounds as a hint that they were once agents of infection (a few are related to known pathogens). Others grow within a membrane that protects them from attack, or make poisons that suppress a host's ability to fight back. The plants have stayed suspicious of their partners. Now and again a cheat gets in – an invader that produces little of its valuable product but demands free food and shelter. At once the host cuts off supplies, the nodule withers and the fraudster starves.

Nature's market in nitrogen turns over billions of tons of the element each year, which passes from air to soil, from land to water and from plants to animals and back again in an endless cycle. As is true for all markets the accounts of profit and loss are checked with great care. The struggle for the element is pitiless as is that for water, air or sex, but only now and again is the truth of its dealings exposed in all its brutishness. Plants that eat animals are just one instance among many to show how competitive that business must be and how the most improbable expedients are pressed into service to squeeze the most out of what little is on offer.

Now the global trade in nitrogen has been thrown into turmoil. Farmers pour nitrogenous chemicals on to the soil. They buy it from factories that each year generate a hundred million tons of the stuff from oil, or by extracting the gas from the air. The reaction is carried out with the help of catalysts in boilers held at high temperature and extreme pressure. Without that technology, invented just a century ago, the world would starve. The industry is profligate indeed in its use of power, most of it gained from burning the remains of ancient life. Cars, chimneys and aircraft also pump nitrogen salts into the air. All this means that far more nitrogen is available in useful form than in Victorian days. The amount has doubled in the past century.

To add fertiliser to fields does increase the yield of crops but also changes the economics of their bargain with a living source of nitrogen. First, it alters the balance of profit and loss. After a dose of fertiliser, crops need less help from their tiny assistants and squeeze them out. As a result the amount of the element taken from the air by those useful creatures goes down, so that the overall gain from the added nitrogen is less than it might otherwise be. For the starved soils of Africa such opportunistic behaviour by the plants is a real problem.

In addition, excess nitrate is washed to where it is not wanted, and more is added by acid rain, itself full of salts of the element emitted by exhausts and chimneys. Insectivores, ant-shelterers and bacterial hosts all respond, for now they have a cheaper source of the crucial nutriment than they did before. The rain-fed bogs of New England were once full of pitcher plants that flourished as they sucked up nitrogen from their prey. Their competitors could not manage in such starved places. The acid marshes have been enriched. In those hardest hit − near cities or close to fertilised fields – the insect-eaters have abandoned their carnivorous habits in favour of a conventional life. Other species move in and drive the pitchers and Venus flytraps to extinction, and in Europe the sundew faces the same problem, which means that the insectivores are converging in death, as they did in life.

Carnivory, which began with shortage, may perish with excess and insects at least can breathe a sigh of relief. To an evolutionist, the shared fate of nitrogen-fixing bacteria and fungi, of the Venus flytrap and the sundew, and of trees and their ants, is further proof, as their diverse talents disappear, that under natural selection, and in both life and death, parallel lines may converge.

CHAPTER III

SHOCK AND AWE

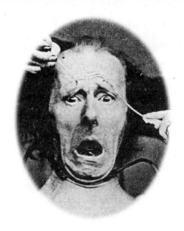

Many American politicians have taken pleasure in gloating over the fact that Zacarias Moussaoui, the Frenchman accused of involvement in the Twin Towers disaster, will certainly go mad, held as he is in solitary confinement in the Colorado 'Supermax' prison. As the judge who passed sentence said: 'You will never get a chance to speak again . . . and will die with a whimper.'

The eminent jurist was not quite justified in his satisfaction at his captive's fate, for many of the tens of thousands kept in endless isolation in American prisons end their lives not with a whimper, but a scream. Some do fall into insanity in such places, but much as the religious right might celebrate their mental decay, they would be dismayed to learn that Moussaoui will lose his mind for Darwinian reasons. Guy the Gorilla, star of London Zoo in the 1950s, was admired for his solemn disposition. In fact, the animal was deeply depressed, kept as he was for years alone in a small cage (although unlike his human equivalents he had no opportunity for suicide). *Homo sapiens* is a social primate and – like gorillas or chimpanzees – descends from an ancestor with the same habits. Had our forefathers been solitary beasts like the orang-utan, which spends most of the year alone, the worst of all punishments would not be solitary confinement but an endless dinner party. The constant exchange of subtle emotional cues around the table would drive all those present to their wits' end.

Science is often asked to explain what makes men different from chimpanzees or orangs but in many ways that is not a scientific issue at all. Such questions deal with the mind rather than the body or the brain; a topic that most competent biologists consider to be outside their expertise. Even so, as scientists compare man's anatomy and behaviour with that of his relatives, biology does reveal a little about how humans became what they are. We are, says all the evidence, creatures that crave society. To satisfy that yearning we spend large parts of our time in silent and sometimes subliminal conversation with each other.

Rousseau had a different view of the origin of human nature. He saw man as in decline from a pure and animal state and modern society as a corruption of what the world should be. 'Savage man, left by Nature to bare instinct alone . . . will begin with purely animal functions . . . His desires do not exceed his physical needs: the only goods he knows in the Universe are food, a female, and rest.' The true life was near-solitude, on a remote island best of all, with a bare minimum of interaction with others. The French philosopher's ideas were romantic, but wrong. Members of all communities, human or otherwise, must negotiate to maintain peace, to have sex and to reap the benefits of cooperation. They use signals both self-evident and subtle to test the mental state of their fellows and to advertise their own, and even the solitary orang hoots now and again to impress its neighbours. Civilisation is based on the ability to respond to another's sentiments and to express a mood of one's own.

In 1879, at the Derby, Darwin's cousin Francis Galton noted that he could assess 'the average tint of the complexion of the British upper classes' as he observed the crowd through his opera-glass. Then the race started, and in a letter to *Nature* entitled 'The Average Flush of Excitement', he observed that it became 'suffused with a strong pink tint, just as though a sun-set glow had fallen upon it'. A shared hue was a statement of a common passion and Galton could work out what it was even when he could not tell individuals apart. In the same way, someone exposed to an image of a group of people who bear a range of expressions from happy to miserable can sense their general state of mind far faster than he could by scanning each visage separately. Our brain, it seems, has a filter that picks up not just how many are in a crowd, but how, on average, they feel. The ability has its down-side. Mass hysteria can spread through society as shared sentiments feed on themselves; as Charles Mackay put it in his 1841 book *Extraordinary Popular Delusions and the Madness of Crowds*, in an account of the South Sea Bubble and other mass fantasies, men 'go mad in herds, while they only recover their senses slowly, and one by one'.

In 1872, in *The Expression of the Emotions in Man and Other Animals*, Darwin discussed the role of signals in the herds, packs, flocks, towns and cities in which social animals come together. The book was a first attempt to understand our own sentiments in scientific terms. He was interested in how mental actions are manifest in the face and the body and realised how closely the displays of inner feeling made by men and women resemble those of animals. The book discusses instinct, learning and reflexes in creatures as different as moths and apes. Its author knew that elephants wept and hippopotami sweated with pain and when he heard a cow grind her jaws in agony he was reminded of the gnashing of teeth in hell. He saw that loneliness, fear or anger and their outer signs have all – like limbs or eyes – evolved. Kick a dog and it crouches and turns down the corners of its mouth; torture an al-Qaeda suspect and he does the same. *The Expression of the Emotions* makes a powerful case for the shared mental descent of humans, primates, dogs and more.

Our own sentiments have long been compared to those of other creatures. The seventeenth-century painter Charles Le Brun, who is referred to in the *Emotions* book as a pioneer in the study of human feelings, urged those who tried to portray their subject's mood to scrutinise beasts first. A few hours with swine, lascivious, gluttonous and lazy as they were, would, he was sure, help depict the inner life of a debauchee. Charles Darwin's friend George Romanes went further. He set out a scale with fifty ranks. Worms and insects came in at step 18 as they could experience surprise and fear; dogs and apes were equal at point 28 as each had 'indefinite morality along with the capacity to experience shame, remorse, deceit and the ludicrous'. Levels 29 to 50 were reserved for men or women of greater or lesser virtue.

Psychology is still marked by such ideas. *Emotions'* central theme was, as ever, a world in which all of life's attributes, from anatomy to anguish, emerge from shared descent. Science uses that logic on elephants, cows, apes, fruit flies and bacteria in its attempts to build a shared narrative of inner feelings. Those who transmit their sentiments expect a response from those who receive them. That two-way commerce involves a need to acknowledge, to copy and to respond to the moods of others. People gasp in sympathy at a sad tale, gaze at where another person's eyes are directed or avoid food that someone else has rejected. Such reflections of another individual's mental state are part of what makes us human.

Charles Darwin, a practical man, had little interest in philosophy.

Even so, he realised that the biology of the mind was harder to interpret than was that of the body. He wrestled with the issue in rather the same way as modern psychologists try to come to grips with some of their own sometimes murky ideas. Can our thoughts be explained just as the 'direct action of the excited nervous system on the body, independently of the will' and if so, what (if anything) does that mean? Shakespeare speaks of Cardinal Wolsey when 'Some strange commotion/ Is in his brain; he bites his lip and starts;/ Stops on a sudden, looks upon the ground . . .' That, Darwin writes, came from the 'undirected overflow of nerve-force' – but is that phrase just an attempt to avoid deeper and less tractable questions? The task was made harder by his quarrel with the anti-evolutionist Charles Bell, author of the standard text on facial anatomy. Bell was convinced – and he was wrong – that humans had unique muscles divinely designed to express morality, spirituality or shame: a notion not of much help to someone anxious to understand the smile or the blush, but an early example of the preconceived truths that still plague many attempts to understand the human mind.

After a long stumble through the Freudian fog, the study of the mental universe has once again become a science, even if the many claims to have found the neural foundations of society do not yet deserve that status. Now, physicists and chemists busy themselves with questions once raised only by intellectuals. In institutes of psychiatry and neurology, cats, mice and dogs are used to dissect human habits. Even bacteria behave in a rational fashion when they settle down close to a source of food, or join hands with their colleagues to form a sticky film over teeth or wounds. Certain fruit-fly genes lead to homosexual behaviour and others to loss of memory, which might one day help in the study of illnesses such as Alzheimer's disease. In mice and monkeys, experiments on brains once done with a scalpel are now carried out with machines of fantastic complexity. They

are also used on people with brains damaged by strokes or accidents, while drugs help understand the mental universe of the
normal, the reckless and the insane. Many of the questions raised
in *The Expression of the Emotions* have a notably modern air and
many remain unanswered.

Emotions is in some ways a less satisfactory work than are the
plant, barnacle or earthworm books and an unusual note of apol-
ogy creeps in: 'Our present subject is very obscure . . . and it
always is advisable to perceive clearly our ignorance' (and there its
author was franker than some of his successors). Charles Darwin
soon found that even what looked simple – the objective description of the facial expression of a man or a dog, for example – was
hard, while to represent the sentiments behind it was even harder.
That problem, in spite of the wonders of electronics, still baffles
students of the nervous system. He was suspicious of phrenology –
the notion that particular segments of the brain are associated
with, for example, obstinacy, pride or guile – even if an admirer
had claimed that the naturalist's own head had 'the bump of
reverence developed enough for ten priests'. He struggled long
and hard with the question as to just where felt experiences
might be seated.

The student of the inner world looked first at the animals and
children of his own household. As a kind-hearted man, he was
careful not to disturb them too much, although his book does
contain images of frightened babies that would see him accused
of cruelty today. His sons, he noted, never pouted, although
Francis's mouth assumed that expression when he played the
flute. He did not hesitate to play the animal himself. Francis
remembered that his father's body was very hairy, and that the
great man would growl like a bear when his children put their
hands inside his shirt.

Even in play the *Beagle's* naturalist was serious, and he soon
identified some general rules about human and animal behaviour.

Intimations of happiness or grief, of welcome or rejection and of other opposed sentiments often came as mirror images. Thus, a frown is the opposite of a smile and a look of surprise the converse of a greeting. Some gestures emerged from movements that once had a function of their own. To beg with open hands is related to the posture taken when holding food and, in the same way, a person who rejects an advance closes his eyes and looks away, as if from an unpalatable meal. Animals seemed to follow similar rules and the paterfamilias of Down House saw almost the same downcast looks in his household pets as those adopted by his infant son.

From such simple observations emerged the science of comparative psychology. It began with dogs.

Pets gain their status because they seem, to their owners at least, to be almost human. Darwin was no exception and kept a dog – Sappho by name – even when he was a student. He saw no problem in describing canine sentiments in the same terms as our own. His pet when in 'a humble and affectionate frame of mind' acted in a way quite different from that of a hostile animal with its bristling hair and stiff gait. The 'principle of antithesis' was hard at work, for opposed sets of muscles were set into action to express contrasting emotions. The 'piteous, hopeless dejection' of his favourite hound when it discovered that it was not about to go out for a walk but instead was to sit in on an experiment in the greenhouse was manifest in a 'hothouse face', the 'head drooping much, the whole body sinking a little and remaining motionless; the ears and tail falling suddenly down, the tail by no means wagged'. That was quite different from its expression when happy and excited, with the head raised, ears erect and tail aloft.

As well as such individual shifts of mood the proud pet-owner noted marked differences in personality among breeds. Descent with modification could, it appeared, change minds as easily as it could bodies. Certain kinds, such as the terrier, grinned when pleased while others did not. Spitz-dogs – huskies, elkhounds and the like – barked while the greyhound was silent. The canine universe encompasses a wide range of talents. Some varieties herd sheep and cattle (and, in the case of the Portuguese Water Dog, chivvy fish instead) while others guard, hunt, guide or annoy the general public. The various breeds when taken together show a wider range of behaviour than that found among all wild canines – wolves, foxes, coyotes and jackals – across the world. Many of the differences are innate, and *The Origin* tells of a cross with a greyhound which gave a family of shepherd dogs a tendency to hunt hares. So impressed was its author with the animals' divergence in habits that he suggested some of the household types had descended from distinct wild ancestors (and there he was wrong).

His favourite pet is back at the centre of the emotional stage. The world has four hundred million dogs and the efforts of their owners and the wonders of science have transformed the creatures into a gigantic experiment on the biology of sentiment. Even in the brief period since modern breeds began to emerge in Victorian times dogs have undergone large – and inherited – changes in temperament.

Men long ago began to use dogs in the hunt. They soon learned to choose those with their own special abilities – to track, to run, to squat into a 'point' position when prey is spotted, or to bite and tear or recover corpses – as parents for the next generation. Such remnants of the chase live on in the behaviour of Bloodhounds, Pointers, Setters, Retrievers and Bull Terriers. Herding dogs such as Border Collies stalk a sheep and do not bite it, but those used to control larger animals – like the Corgis once used with cattle – go further through the sequence and snap at their charges. Pit Bulls complete the job and are vicious creatures that will hold a bull by the nose and as a side-effect sometimes kill their owners. Guard dogs such as Pyrenean Mountain Dogs, whose job is to frighten off predators, have given up the hunt sequence altogether. They play like huge puppies and show little

interest in their herds, but their conduct is odd enough to persuade wolves to stay away. Such differences emerge from inherited
variation in behaviour within the common ancestors of each
breed, from new genetic errors as the generations succeed each
other, and from the accumulation of change by human choice.

One way to assess a dog's personality is to startle it with the
appearance of a stranger. Does the animal play with the visitor,
back off, sniff him or chase him out of the room? Does a
sudden noise anger the beast, terrify it or leave it unmoved?
Other tests include the ability to sit still, to cope with solitude
without whining or panic, to run through a maze or to find
hidden food.

Cocker Spaniels are calm and obey orders, while Basenjis are
nervous and almost impossible to train. Crosses between the two
suggest that the difference in their nature is inborn, for the offspring have a range of talents, intermediate between each parent.
A survey of ten thousand German Shepherds and Rottweilers in
Sweden showed, within each type, a shared inheritance of
excitability, a tendency to wag the tail and a need to bark, while
aggression appears to be under separate control. In an echo of
Expression's principle that antithetical emotions are expressed as
mirror images, variation in all those capacities depends on just
how shy or how bold a particular breed might be.

As the dog-fanciers' tastes became more refined, more and
more specialised varieties emerged. Some began to develop habits
that perturbed their owners. Mating like with like exposed rare
and once-hidden genes, many of which had undesirable effects on
personality. Some have parallels in the mental lives of men and
women. In an echo of human obsessive-compulsive disorder, Bull
Terriers chase their own tails for hours until they collapse, while
Springer Spaniels may savage their masters as they fall into a
sudden attack of uncontrolled rage. Certain families of Bassett
Hounds suffer from a delusion reminiscent of paranoid schizophrenia and cower at the slightest noise. Some Dobermans, in

contrast, fall into a heavy slumber after an unexpected snack. They have narcolepsy, a distressing and sometimes dangerous condition also found in people – and the dogs respond well to the drugs used to treat human patients.

The double helix reveals why some breeds diverge so much in personality. The first complete sequence came from a Boxer. The animal had less DNA than we do, with about twenty thousand genes altogether, several thousand fewer than ourselves. The hope is to find canine matches to our own disorders, and some have already emerged. The sleep problem in Dobermans involves damage to a certain receptor protein on the surface of brain cells – and the human equivalent is due to a fault in the same gene. No doubt our companions will help track down many more of the inherited errors behind our own mental illnesses, as they already have for conditions such as blindness. Charles Darwin would be proud.

Dogs are anomalous animals for their habits have been so subdivided by human effort that their mental universe is far from typical of a wild creature. Darwin soon moved on in his search for the roots of human emotion. He spent many hours in the company of our relatives in London Zoo. He had particular fun with the anatomy of amusement: 'Young Orangs, when tickled, likewise grin and make a chuckling sound . . . as soon as their laughter ceases, an expression may be detected passing over their faces, which, as Mr. Wallace remarked to me, may be called a smile . . . I tickled the nose of a chimpanzee with a straw, and as it crumpled up its face, slight vertical furrows appeared between the eyebrows. I have never seen a frown on the forehead of the orang.' He was particularly taken by the attempts of a monkey to court its own image in a mirror and by the antics of Jenny the orang-utan, who when teased with an apple on the wrong side of the bars 'threw herself on her back, kicked and cried, precisely like a naughty child'.

Primates, like people, reveal their feelings on their faces. Someone who has never before seen a macaque can at once identify its
mood as sad, happy or enraged, when shown the appropriate
photograph. Many chimpanzee expressions have been named.
They include the closed-mouth smile, its bared-teeth equivalent
(which descends from an ancestor shared with our own smile),
the bozo smile and the play face (a relative of human laughter),
together with subtler statements of mood such as the stretched
pout-whimper. Bonobos have an amused expression – and noise –
which is uncannily like a guffaw. A German expert has identified
an *Orgasmusgesicht* or 'orgasm countenance' in that species,
although its existence in humans remains to be demonstrated.
Gorillas are more impassive for they grin and make bozo faces
but otherwise keep their thoughts to themselves unless they are
simply furious.

Apes and monkeys can interpret their fellows' moods to a considerable degree. Electronic avatars of chimps can have their looks
manipulated to simulate pout-whimpers and the rest. When real
animals are presented with their artificial comrades, they pick out
different expressions at once, screaming faces best of all. They
also show some insight into another's emotions. If one animal
sees another grimace in fear when it hears a sound it had learned
to associate with an electrical shock, the observer will flinch
when the buzzer goes off, even if it has never itself experienced
the shock.

Humans are even better at sensing the moods of others. We are
so aware of facial features that we often see them when they are
not there (which explains the sad ape-like countenance in NASA's
pictures of the mountains on Mars). Two scrunched-up newspapers look much the same although their shapes are quite different,
while two faces are seen as quite distinct although their shapes are
almost the same. A simple bar code, the position of six stripes of
dark and light – hair, forehead, eyebrows, nose, lips and chin –
stores most of the data. Most people can recognise thousands of

individuals and sense dozens of emotional states. Faces are important even to infants. Darwin noted that, when they were very
small, his children spent long periods gazing at their mother. Now
we know that a baby responds to a human countenance – even in
a photograph – within minutes of birth.

Men, like apes, speak with their faces and use more or less the
same language to do so. Angry people and angry gorillas bare their
teeth and a frightened chimpanzee looks rather like a frightened
person. For humans, as for apes, some expressions are ambiguous.
Men and apes bare their teeth when amused but do the same when
filled with terror. *Emotions* has a picture of a Sulawesi macaque as
it grins in pleasure when stroked – but in other macaques the same
gesture marks submission to a threat. Not all our grimaces are
shared with our relatives, for apes never signal disgust and their
noses, which are more sensitive than our own, remain unwrinkled
even to a repulsive smell. A wide-open mouth is a threat in many
primates but conveys no more than mild surprise for humans and
while elephants weep, our closest kin do not.

Monkeys and apes reflect their moods in their postures as well
as their expressions, and gorillas really do slap their chests in rage.
Men, orangs, chimps and gorillas share the Italianate habit of
waving their hands. Bonobos flap their wrists in irritation, point
at themselves when they need a hug and stick out their palms
when food is on offer. In a further nod at our common heritage,
they prefer to signal with the right hand.

Our faces are more eloquent than are those of any other primate.
Many pages of *The Expression of the Emotions* are devoted to the
way they reflect their owners' inner state. Some read rather quaintly
nowadays: 'the breach of the laws of etiquette, that is, any impolite-
ness or *gaucherie*, any impropriety, or an inappropriate remark,
though quite accidental, will cause the most intense blushing of
which a man is capable. Even the recollection of such an act, after
an interval of many years, will make the whole body to tingle. So

strong, also, is the power of sympathy that a sensitive person, as a lady has assured me, will sometimes blush at a flagrant breach of etiquette by a perfect stranger, though the act may in no way concern her.' In the interests of science, modesty gave way to the search for truth: 'Moreau gives a detailed account of the blushing of a Madagascar negress-slave when forced by her brutal master to exhibit her naked bosom' and the sexual nature of that expression means that 'Circassian women who are capable of blushing, invariably fetch a higher price in the seraglio of the Sultan.' Mark Twain, himself an ardent evolutionist, put it well: 'Man is the Animal that Blushes. He is the only one that does it—or has occasion to.'

Darwin was keen to discover whether signals such as the blush were the same in every human culture, or whether, like skin colour, they changed from place to place. He rejected the popular notion that different races had evolved from higher or lower forms of primate and that their mental lives and expressions of mood reflected this. Soon he began to accumulate a mass of anecdotes that made the case for the universal nature of facial cues. People also wrote to him about their dogs frowning in concentration or showing moral courage when their teeth were pulled. His correspondents included William Ewart Gladstone, who commented on statements about the Greek visage found in Homer, but also 'Captain Speedy who long resided with the Abyssinians; Mr Bridges, a catechist residing with the Fuegians and Mr Archibald O. Lang of Coranderik, Victoria, a teacher at a school where aborigines, old and young, are collected from all parts of the colony'. One letter told of a Bengali boy with 'a thoroughly canine snarl'. Its recipient fired off a series of questions to those servants of the Queen, sometimes to ludicrous effect ('Mr B.F. Hartshorne . . . states in the most positive manner that the Weddas of Ceylon never laugh. Every conceivable incitive to laughter was used in vain. When asked whether they ever laughed, they replied: "No, what is there to laugh at?"').

In spite of the Weddas, Darwin became certain that such signs were more or less universal across the globe: 'The young and the old of widely different races, both with man and animals, express the same state of mind by the same movements.' Hard as it is to believe, that observation was forgotten and for many years students of humankind assumed that expressions were determined by culture and were not coded into DNA (even if nobody found a place where people laughed in pain or screamed in welcome). Looks of anger, disgust, contempt, fear, joy, sadness and surprise all are universal. One tribe in New Guinea cannot separate expressions of fear from those of surprise – but in that society any intruder is a threat. People from different cultures do find it harder to identify each other's guilty or shamefaced looks than they do a smile or an expression of terror, so that such subtle statements of mood may in part be learned. Even smiles are equivocal, for the beam, grin, smirk, snigger, simper and leer each convey a different message while people who smile too often come across as nervous rather than contented. Darwin, too, saw some ambiguities. The expression in a photograph of a man almost in tears was recognised by some as a 'cunning leer', a 'jocund' frame of mind or even as someone 'looking at a distant object'.

Once he had established that most such signs were common to all mankind, Darwin set out to describe them. Measurement, he knew, is the first step in science (a lesson still ignored on the wilder shores of psychology) and he tried hard to give an impartial description of human features ('The contraction of this muscle draws downwards and outwards the corners of the mouth, including the outer part of the upper lip . . . the commissure or line of junction of the two lips forms a curved line with the concavity downwards and the lips themselves are generally somewhat protruded').

In today's world of fraud, terrorism and identity cards such attempts to put facts on to faces have become an industry.

Remarkable claims are made about the ability to identify people and to sense their states of mind. Some enthusiasts recognise thirty indications of anger and eight of sadness, with additional criteria based on how the subject holds his head. George W. Bush's countenance was more or less blank whatever message he tried to put across, but his Department of Homeland Security has spent millions on machines that claim to detect when a terrorist is about to attack by the look on his countenance. Nobody denies that the expression of a Scotsman with a grievance is easy to distinguish from a ray of sunshine but such claims go too far.

The face says a lot about how we feel, but – as in apes – the body adds information to the stream of emotional cues: a man with raised fists is not about to make a visitor welcome. Psychologists tend, for practical reasons, to use pictures of faces alone. That can be a mistake. An image of a man with a disgusted expression taken from a modern catalogue of facial poses is interpreted as manifesting revulsion when superimposed on to a body holding a pair of dirty underpants – but as a look of anger when added to a torso with fists raised, or of triumph when stuck on to the beefy frame of a body-builder. The same photograph shown against the background of a cemetery is interpreted in a different way than when seen against a neutral surface. For students of the emotions, the assumption of simplicity can confuse results taken from the most complicated machines.

The face is a real mirror to the soul. Even a brief glimpse reveals the presence of another person, identifies who it might be and gives a strong hint as to what its bearer will do next. Most westerners interpret a set of features with a quick triangular scan of both eyes and the mouth, each of which say a lot about identity and state of mind – but the Chinese tend to concentrate their attentions instead in a fixed look at the nose and pick up the general expression of the whole visage in the background. Scans show that when someone flashes into view, the brain first notes his or her presence, then identifies who it might be and last of all of tests their mood: this is a face, it belongs to Fred, Fred is furious. It processes a portrait twice as fast as a picture of other objects. A certain part lights up about a tenth of a second after a face is first seen, notes its identity about a fifteenth of a second later and takes even longer to interpret what humour the person might be in.

Some expressions are easier to identify than are others. The smile is coded deep within the skull and everyone has an inborn ability to assume it. As Darwin noted, babies born blind smile without difficulty and (as he did not) blind athletes raise their arms in the air in a chimp-like gesture of triumph when they win. Children find it easier to pick out expressions of good cheer than they do those of fear or disgust. Women smile more at strangers than do men, while men are worse at working out mood from a slight movement of the lips. A lopsided grin to the right is seen by most of us as more joyful than is its equivalent on the left. Even sheep, when given a choice of a smirking or a sombre shepherd from whom to take food, prefer the cheery individual. We smile or raise our arms not to reassure ourselves that we are happy or proud, but to tell others how we feel. Context is all; when Chelsea score, fans respond with roars of triumph rather than smiles of delight, but gold medal winners as they stand on the Olympic podium have wider grins than do those who have gained bronze.

Signs of delight or terror seem simple enough, but there are real differences in the ability to decode them. I have a talent that illustrates that fact, for I can waggle my eyebrows. It began in school when I was rebuked for glowering. I then tried dumb insolence with a one-brow grimace rather than the full two-brow scowl and in time it became easy to alternate. It is still my occasional habit to amuse small children with the trick – and almost always they smile back. Unfortunately, an occasional infant screams instead. The signal is clear but the response uncertain.

Both steps can go wrong. Some people cannot tell individuals apart from their faces and use clues from voice or clothes instead. In one instance, a litigant wandered into court and discussed his case with a barrister – not his own, but his opponent's. The context was right; a lawyer, with a gown, in a courtroom. The face alone did not fit. Needless to say, he lost. Face-blindness may be caused by a stroke, but a certain form runs in families with perhaps just a single gene involved.

Other unfortunates lose the ability to broadcast their emotions. For some reason – injury, infection, cancer or brain haemorrhage – the facial nerve no longer works and the patients cannot express their feelings. They find it hard to assume looks of happiness, fear or surprise, and their wives, husbands and friends soon notice the problem. The condition might appear to be trivial, but in fact causes real distress and sometimes even suicide, most of all when an attempt to smile emerges as a grimace or a leer because the eyebrows – usually lifted at a happy moment – refuse to obey instructions. Some people have their brows surgically moved upwards (which gives them a permanent look of surprise), while others grow a long fringe that hides the offending forehead. The readiness to take such steps shows how much a signal of mood is a passport to society.

Emotions marked the first real attempt by science to infer the action of the mind from its external signs. Scientists now study the activity of brain cells rather than of facial muscles as they try to understand our inner feelings. The use of electricity – and of the sophisticated electronic devices that depend on it – in psychology has become a science of its own. It was first expounded in Charles Darwin's book.

The ancient Greeks had used electric fish to treat headaches but for many years the galvanic fluid was no more than an entertainment. An entire community of monks was once connected by a mile-long iron wire and made to jump for the amusement of the

King of France (castrati were tested to see if they acted as insulators, but they did not). *Emotions* contains several pictures of faces stimulated by shocks to give expressions that resemble the natural look of horror, rage and the like. They came from the French physician Guillaume-Benjamin-Armand Duchenne de Boulogne. Duchenne is best remembered for the muscle disease named after him but he also studied the expression of what he called the 'passions', using electrodes touched to different parts of a countenance to stimulate the muscles. He was the first to notice that a genuine smile involved raised eyebrows, and his machine could easily activate those 'sweet muscles of the soul' to simulate a happy beam. He even made the visage of a decapitated criminal assume a simulacrum of pleasure with a probe upon its cheek. Duchenne chose as his main subject an aged man of feeble intellect, for he 'wanted to prove that, despite defects of shape and lack of plastic beauty, every human face can become spiritually beautiful through the accurate rendering of emotions'. His pictures first came to public attention when they were published in *The Expression of the Emotions*. They played an important part in Darwin's attempts to give an objective account of expressions of pleasure or pain.

The machines have marched on. Where Duchenne used a battery, a metal rod and a plate-camera, scientists in search of the springs of sentiment now depend on electro-encephalograms, positron emission tomography or functional magnetic resonance imaging (FMRI). Tiny electrodes are used to activate single nerve cells, while the EEG and its relative the magneto-encephalograph pick up electrical activity within the skull. PET scanners use a sugar marked with a radioactive label which is taken up by active parts of the brain and then detect its decay products. The fMRI machine, in contrast, senses tiny changes of blood flow through the grey matter from a shift in the magnetic properties of the red pigment haemoglobin as it gains or loses oxygen.

Marvellous as such techniques are, they run into many of the problems that plagued Charles Darwin. He had found it hard to decide just where the jaw ends and the cheek begins or to identify the precise arrangement of facial muscles. Today's arguments about the boundaries between areas of the brain as defined by electronic scans – confidently coloured and labelled as the images might be – reflect his own doubts about the anatomy of the human countenance. Some claim that particular emotions can be mapped to a definite part of that organ. Others see the brain – as he saw the face – as a connected structure, with most sections contributing to most of its functions. Any attempt to pinpoint centres of anger, joy or despair might be of its nature a mistake.

Another problem for both the nineteenth and the twenty-first centuries comes from the need to describe broad sentiments in narrow terms. Darwin was happy to talk about dogs in a 'humble and affectionate frame' of mind – but how is it possible to put figures on humility or affection? Objective fact soon slides into mere interpretation and *Expression* was itself not immune from that temptation. Its photographic plates are not originals but engravings, some touched up to make a point. A mad lady with tousled hair was given a furrowed brow by the engraver and a screaming infant was made to look even more miserable than before by copying the portrait and re-photographing the sketch (the picture sold hundreds of thousands of copies to a gullible public). The stream of lurid images of centres for pain, passion and pleasure that decorate the scientific literature and leak into the press are also in some senses fakes. Digital information is processed in a complicated and sometimes subjective way to make a picture which is often rather more than the sum of its parts.

A final difficulty for both the Victorians and their descendants was to find subjects who were willing to display their emotions to the world. Duchenne set up a theatre in which the public could be delighted by actors galvanically activated to produce an air of grief or delight. Many of *Expression*'s pictures are also based on members of that profession. Among them, a bearded thespian looks remarkably implausible as he strikes his attitudes. Actors still play an important part in neuroscience. Their photographs are taken as they simulate a mood and are then shown to subjects whose brains are scanned to see which bits light up. Many of the images look just as posed as do those of Darwin's theatrical friend. The artistes over-do the job, often to a bizarre degree. People shown pictures of frightened or unhappy people taken from real life have far less of a nervous response than they do to images of those who simulate a mood as they strut and fret upon the laboratory floor. Most of us find it harder to interpret the sentiments (apart from laughter) in silent clips of Hollywood stars from Dustin Hoffman to Meryl Streep than we do the simulated joy or terror of a ham actor – and yet the hams are used as raw material for experiments of huge technical sophistication and expense.

Many claims have been made that particular parts of the brain respond to the sight of a happy or miserable countenance and that they prepare the nervous system to beam back or look sympathetic in return, but they have been hard to replicate. Because light comedy is a subtler form of entertainment than is Greek tragedy, the scientists who study that great theatre of emotion, the face, often focus instead not on mild signs of contentment or sadness but on expressions of horror and dread that might provoke an unambiguous response in those who see them.

A blank stare is a signal of terror and Shakespeare knew as much. A furious Othello says to his supposedly unfaithful (and frightened) wife Desdemona before he kills her: 'Let me see your eyes.' We have larger eye-whites than any other primate and take more notice of them, for the mouth is far more important than the eyes as a chimpanzee emotional signal. We process eyes quicker than any other feature and fearful, stretched-open eyes even faster – and women do the job better than men. One woman could not recognise a picture of a terrified individual because she did not look at the eyes. When instructed to do so she at once understood the subject's frame of mind.

The brain's main activity in response to a frightened look takes place in a pair of structures called the amygdalae. They are almond-shaped groups of nerve cells deep within the temporal lobes, the side sections of the brain, one on each side, embedded into what is sometimes seen as the organ's most primitive parts. Each is connected to other brain centres, to the hypothalamus – that hormonal bridge between the nervous system and the blood-stream – to nerves that feed from pain receptors and from the eyes, and, in primates more than other mammals, to nerves to and from the face itself.

Animals in which the structures have been damaged find it hard to pass the classic test in which fear of an electric shock becomes associated with the sound of a bell. Experiments on monkeys in which those parts of the brain were cut out showed that the unfortunate creatures in addition lost their ability to recognise familiar objects, together with their nervousness about humans, and a mother's affection for her infant. Human patients with dam-aged amygdalae have similar problems with emotionally draining tasks. The amygdala is also involved in memory. People recall where they were on 11 September 2001 with its help, but those in whom the structure is damaged remember the Twin Towers disaster no better than what they had for breakfast.

The amygdalae are busiest when a frightened gaze is directed straight at its target – which fits Darwin's idea that a countenance stricken by terror is an immediate signal of danger. A few people have such severe brain damage that they perceive themselves as blind – but show them a scared person and the amygdala lights up. We are slower to notice the racial origin of an angry than of a happy face, so that fear has priority over familiarity. In the United States, images of black people shown to whites stir up more activ-ity than do those of individuals of their own skin colour.

The case for the amygdala looks persuasive but, as usual when it comes to the contents of the skull, real life is not simple. Other parts of the brain are also involved in the response to a terrified countenance. The amygdala lights up in response to a whole face rather than just the eyes, and does so to some degree whether or not the subject shows signs of alarm. Its main role might be to notice new events, whatever they might be, rather than to make a specific response to a particular emotion.

The structure helps to process a nerve-transmitter called serotonin (which is also involved in temperature control, sleep, hunger, lust, response to injury, liver repair and more). Many antidepressants work because they change the way in which serotonin is broken down, or taken into cells. Variation in the ability to respond to or to make the substance might be behind individual responses to fear. Some people are terrified even by the simplest problems of society. Darwin writes of a dinner party given for a man who, in response, 'did not utter a single word; but he acted as if he were speaking with much emphasis. His friends, perceiving how the case stood, loudly applauded the imaginary bursts of eloquence, whenever his gestures indicated a pause, and the man never discovered that he had remained the whole time completely silent.' The unfortunate fellow could now be comforted with the information that he may have a more active amygdala than normal and that his nervousness might be treated with drugs that alter his body chemistry.

Inborn errors in the ability to synthesise serotonin make some people sad, angry or suicidal. A gene whose product helps remove the chemical from the junctions between nerve cells comes in two common forms, one better at the job than the other. The less active type is more frequent among people who are anxious, neurotic or depressed – and its bearers are less able to decode expressions of fear or sadness than are their fellows. The orangutan – the most solitary of our primate kin – has a version of the gene that is even less busy than that of the most socially isolated human. Whether its feeble serotonin pump has much to do with its lonely life and presumed dislike of dinner parties remains to be proved.

People with severe depression often find it hard to sense the emotions of others. Drugs that affect serotonin can help the illness – and their immediate effect, sometimes within hours of the first pill, is to improve a patient's ability to interpret their fellow citizens' feelings from their faces. That simple talent turns the key that restores them to society.

Nowhere is the importance of signals better seen than in children. When very young their insights are limited and self-centred, but soon they begin to understand and to respond to the moods of those around them. Darwin wrote a *Biographical Sketch of an Infant*, an account of child development based on his son William: 'When 110 days old he was exceedingly amused by a pinafore being thrown over his face and then suddenly withdrawn; and so he was when I suddenly uncovered my own face and approached his. He then uttered a little noise which was an incipient laugh.' William 'did not spontaneously exhibit affection by overt acts until a little above a year old, namely, by kissing several times his nurse who had been absent for a short time'. By then he could tell faces apart (some of which pleased him more than others) and could copy movements. By the age of eighteen months most children can separate false movements of anger or upset made in play from real gestures and by five – school age – they send and receive information well enough to allow them to live in groups, to learn and, in time, to join society. A sense of self and a sense of other are closely related, for the younger a child is able to recognise a picture of itself the better it interacts with its fellows when it grows up.

William and his brothers and sisters were lucky for they were raised in an affectionate household. Many youngsters are less fortunate. An infant brought up in isolation or by cruel parents may never adjust to the world around it and can feel isolated for the rest of its life. The fit between childhood abuse and adult depression is well established and those taken into care because of poor parenting are at far higher than average risk of emotional problems later in life. A failure to be provided with the signals of affection that bind children to their mothers and fathers and to society as a whole is to blame.

A few unfortunates suffer from loneliness or despair for the opposite reason. What condemns them is not neglect by those who should provide the crucial emotional messages, but their own inability to receive and interpret them. Such children are often diagnosed as autistic. They may live in isolation and unhappiness, with an existence that can seem scarcely human at all, for children with severe autism cannot make or understand the cues needed to find a place among their peers. Their plight shows how central is the ability to express, and to understand, emotions in allowing every citizen to take part in society.

Autistic children are now treated with sympathy and concern, but once they were regarded almost as animals. To those curious about where the essence of humanity might come from, they were useful raw material for speculation. Rousseau wondered whether a youth brought up 'wild, untamable and free' would be safe from the corruption faced by those who undergo a normal education. He pondered an 'impossible experiment': to raise a newborn infant in isolation, but as he wrote, 'by our very study of man, the knowledge of him is put out of our power' – nobody would be so cruel as to do such a thing. Such a child might, he thought, show how the true signals of inner sentiment emerge in a creature that had never received them.

The eighteenth century was a vintage era for 'wild children', those raised – metaphorically or otherwise – by wolves, in the fashion of Romulus and Remus. The naturalist Linnaeus classified them as *Homo ferus* – wild men – whose nature would reveal what made thinking humans, *Homo sapiens*, different. Most of the supposed examples were fakes, but a few were not.

In 1797, a young boy was found alone and almost naked in the forests of the Aveyron, in south-central France. He was captured, escaped, recaptured and escaped again, but in time he emerged from the woods under his own volition. He was about twelve years old, unable to speak and savage in his behaviour. A vicious scar on his throat hinted that his parents had tried, but failed, to kill their aggravating child. The lad appeared to have been without contact with others for almost his whole life and showed no obvious signs of joy, fear or gratitude when at last he met members of his own species. Here, perhaps, was an opportunity to investigate the springs of emotion.

A young student, Jean-Marc-Gaspard Itard, heard the story and saw the chance to test Rousseau's ideas. He took the forlorn boy to Paris and set out to try to raise him to the spiritual level of his fellow citizens.

Itard had trained as a tradesman, but took up medicine at the time of the French Revolution and later became a pioneer in the study of diseases of the ear, nose and throat. In stark contrast to Rousseau he was convinced that the essence of the human condition lay in the ability to sense the feelings of others and, armed with that talent, to build a society in which passions could be kept in check for the good of all. In his 'Historical Account of the Discovery and Education of a Savage Man' he set out his theory that 'MAN can find only in the bosom of society the eminent station that was destined for him in nature . . . that moral superiority which has been said to be natural to man, is merely the result of civilization'.

The doctor took young Victor – whom he named after one of the few sounds, 'o' (as in the French word for water), he was able to recognise – into his household and attempted to train him to express, and respond to, emotions. He was soon disappointed. The boy was 'insensible to every species of moral affection, his discernment was never excited but by the stimulus of gluttony; his pleasure, an agreeable sensation of the organs of taste; his intelligence, a susceptibility of producing incoherent ideas, connected with his physical wants; in a word, his whole existence was a life purely animal'.

Itard laboured for five years with both kindness and cruelty (the latter based on his charge's fear of heights) to transform the boy from monster into Frenchman, but with little success. Victor's behaviour stayed strange: he was obsessed with the sound of cracking walnuts but ignored gunshots close to his ears, and loved to rock water back and forth in a cup. He never learned to speak and showed no gratitude for food or shelter. The sole sign he made of any response to the sentiments of others was that, when Itard's housekeeper was in tears after the death of her husband, Victor appeared to comfort her. Apart from that he stayed apart from his fellow men.

His protector insisted that the young man's failure to adapt to the inner worlds of those around him and to express his own feelings arose because he had been rescued too late to pick up the skills needed, but that view was too optimistic. The lad would nowadays be diagnosed as deeply autistic; as unable to respond to, or give, the signs – the smiles or frowns or conversations – that bind people to their parents, to their friends and to the community in which they live. The dire effects of the illness show how the expression of our own emotions and our response to those of others makes us what we are.

The term 'autism' was invented in the 1940s to describe a condition in which children fail to interact or to smile or express sentiments apart from anger and unhappiness. They speak with difficulty or not at all and are filled with obsessions about particular foods, places or clothes. About a third suffer from epilepsy. Three out of four of those with a grave form of the illness struggle to cope with society throughout their lives. Autism shades from the severe disturbance shown by Itard's Wild Boy himself, through Asperger's syndrome, in which the language problems are less marked, to general problems in the development of normal conduct. Often, the problem is noticed when parents become concerned by their child's depression or rage. Some autists, once unkindly referred to as *idiots savants*, have remarkable talents in drawing or in particular mathematical tasks, but their gifts do no more than disguise their deeper problems. Once, the illness was said to be rare, with one child in two thousand affected, but now the diagnosis is made far more often, with an incidence of one in a hundred in Britain.

Autists cannot understand the signals of their fellows or make the full complement of their own. All children have that difficulty in their earliest years. As Darwin wrote in the *Sketch of an Infant*, 'No one can have attended to very young children without being struck at the unabashed manner in which they fixedly stare without blinking their eyes at a new face; an old person can look in this manner only at an animal or inanimate object. This, I believe, is the result of young children not thinking in the least about themselves, and therefore not being in the least shy, though they are sometimes afraid of strangers.' For most infants such self-absorption soon passes but an autistic child is locked into that phase for life. Many, when they look at other people, ignore the eyes, the flags of sentiment. They are just as unconcerned when someone else gazes long and hard at them.

The Expression of the Emotions used the blush as a prime example of a social cue but embarrassment plays a lesser part in life today. Yawns – unacceptable in a nineteenth-century parlour – are more frequent. We do not know why we open our mouths when tired or bored (although the book discusses the gesture as a threat in baboons). Yawn and the world yawns with you and even to read about it can spark the gesture off, as about half the readers of this book can now attest. The habit begins at about the age of six. Not, however, for children with autism, for a yawn sparks off far fewer responses among them than among the general population. Such failures of empathy lie behind many of their problems.

Psychologists talk of 'theory of mind', the ability to infer the mental state of others from smiles, frowns, gestures and speech. People with autism have little or no insight into the inner world of their fellows and cannot express their own internal universe in a way that makes much sense to those around them. They are blind to the messages written on another's countenance and find it hard to separate gestures of anger, fear, sadness or joy. Like chimpanzees (but unlike dogs) some autistic children cannot understand what is meant when their parent or doctor points at an object. They are denied even that simple social talent.

Autists also find it harder to tell people apart or to recognise a photograph of themselves. A certain group of brain cells is activated when monkeys or men see or copy the movements of others or observe an expression of pain, fear or disgust. They are also involved in the shared response to a yawn or a smile. These mirror neurons, as they are called, are almost silent in children with the severe form of the disease. Perhaps they are part of the system that helps us see into the souls of those around us. In their failure they condemn people with autism to a world whose other denizens act in a mysterious and unpredictable way.

Nobody knows what causes autism and the condition has no cure, even if some of its symptoms such as insomnia or depression can be treated. The illness is four times more frequent in boys than girls, but shows no fit with race, social class or parental education. Infection, immune problems, vaccines, heavy metals, drug use while pregnant, Caesarean births and defective family structure in Freudian mode (the child psychologist Bruno Bettelheim spoke of 'refrigerator mothers') have all been blamed but those claims do not stand up. Some say that the brain of a typical autistic child grows too fast too soon, but then slows down. The amygdalae – those detectors of fear – are overactive in some patients, but many other parts of the brain have also been implicated. Problems with serotonin, that universal alibi for disorders of emotion, may be to blame, for some autistic children synthesise the stuff less well than normal. Certain drugs used against depression can help, as a further hint of a tie between social isolation and the emotional universe.

Genes are without doubt involved in some patients, even if not more than a tenth or so of cases can be ascribed to a definite genetic cause. If an identical twin has autism its sib is at a seven-in-ten risk while the figure risk for non-identicals is far lower. The incidence increases by twenty times above average in the brothers and sisters of those with autism and some among them are tactless, aloof or silent but are not diagnosed as ill.

Such behaviour sometimes presents itself as part of a larger medical problem. Fragile-X syndrome is the commonest cause of mental disability among boys. It comes from a huge multiplication of a short segment of DNA upon the X chromosome. Some patients have symptoms quite like those of autism and some individuals diagnosed with that condition may in fact have the chromosomal abnormality. Other deletions, duplications or reversals of a segment of chromosome are behind other cases of the illness. Often, these arise anew in the children compared with their parents. Some badly affected patients have problems with a gene involved in the transmission of impulses between nerves. A few may have errors elsewhere in the DNA – and dozens of genes, with a variety of tasks, have been blamed. One candidate belongs to a group of genes that is multiplied in number in humans compared with all other mammals, is active in the brain and is damaged in at least a few autistic children. In spite of such hints the biology of autism remains obscure and there are likely to be several explanations for a condition that is not a single disease but many.

Autistic children are an experiment in emotion. Their isolation is mental rather than physical, for they are cut off by an inability to respond to the flow of information that passes between others. A world full of autists could not function, for all societies depend on a silent dialogue in which every member's intentions are overtly or otherwise expressed. Civilisation turns on the ability to bear another's company.

Those who break civilisation's rules must be punished; and part of that invariably involves the manipulation of a criminal's mental state. Prisons are, of their nature, places in which social interactions are forcibly reduced. Solitary confinement is a penalty far more severe than mere imprisonment, for it is autism imposed: a permanent denial of what it means to be human, inflicted upon someone who once experienced the full range of human emotion. The penalty is bitter indeed and is much appealed to by punitive societies, from medieval England to the modern United States. Charles Dickens visited such a penitentiary in Philadelphia and wrote that 'I hold this slow and daily tampering with the mysteries of the brain to be immeasurably worse than any torture of the body.' The infamous 'Supermax' at Marion, Illinois, a jail built to hold the most violent offenders, together with political prisoners such as Black Panthers and members of the American Indian Movement, allowed almost nobody out of their cells for twenty years, even to exercise. It closed in 2007, but some of its forty and more replacements are just as brutal. Some even feed their inmates on tasteless 'Nutraloaf' further to reduce their contact with the world of the senses. Many inmates – like autists – become anxious, agitated and angry, and may end in insanity, killing themselves should the chance arise.

If Zacarias Moussaoui, sentenced to life in solitary for his supposed ties to the Twin Towers outrage, were allowed reading material in his soundproofed Colorado cell he might learn something from both Dickens and Darwin about why he feels such hatred for those who do not share his views. As books are not available, he may wish instead to spend his solitary hours in contemplation of the expression of a condemned prisoner as the electricity passes through his head, which is said – in an echo of the great naturalist's own observations – to be of ungovernable horror.

CHAPTER IV

THE TRIUMPH OF THE WELL-BRED

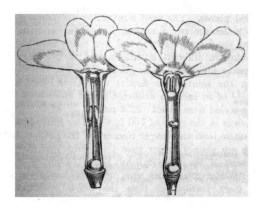

Charles Darwin was worried about his plans for marriage. Perhaps the whole idea was a mistake because of the time that would be wasted on family life at the expense of science. His diary records how he agonised over the pros and cons of matrimony, and his decision to 'Marry, marry, marry!' And marry, in the end, he did.

His spouse was his cousin, Emma Wedgwood. In falling for a relative he stuck to a clan tradition. The Darwins, like many among the Victorian upper crust, had long preferred to share a bed with their kin. Charles's grandfather Josiah Wedgwood set up home with his third cousin Sarah Wedgwood. Their daughter, Susannah, chose Robert Darwin, Charles's father. Charles's uncle – Emma's father – had nine offspring, four of whom married cousins. The evolutionist's own marriage was in the end happy, with ten children (and when his wife was in her forties he wrote that 'Emma has been very neglectful of late for we have not had a child for more than one whole year'). Even so, in Queen Victoria's fecund days the Darwin-Wedgwood dynasty did less well than most, for among the sixty-two uncles, cousins and aunts (Emma and Charles included) who descended from Josiah, thirty-eight had no progeny that survived to adulthood.

Six years after his wife's last confinement Darwin began to think about the dangers of inbreeding, in particular as they applied to his own choice of spouse. His concern was picked up from another of his cousins, Francis Galton, the founder of eugenics, who had pointed out the potential dangers of marriage within the clan.

Charles was anxious about his children. His tenth and last, Charles the younger, died while a baby; he was 'backward in walking & talking, but intelligent and observant'. Henrietta had a digestive illness not unlike that of her father and took to her bed for years, and he feared that his son Leonard was 'rather slow and backward' (which did not prevent his later marriage to his own cousin or his acceptance of the Presidency of the Eugenics Society), while Horace had 'attacks, many times a day, of shuddering & gasping & hysterical sobbing, semi-convulsive movements, with much distress of feeling'. His second daughter, Elizabeth, 'shivers & makes as many extraordinary grimaces as ever'. George's problem was an irregular pulse, which hinted at 'some deep flaw in his constitution' and, worst of all, his beloved Annie expired at the age of ten, throwing her parents into despair. As he wrote, 'When we hear it said that a man carries in his constitution the seeds of an inherited disease there is much literal truth in the expression.' Once he even wrote to a friend that 'We are a wretched family & ought to be exterminated.' Might his own illness and that of his sons and daughters be due to his own and his ancestors' choice of a relative as life-partner? Was inbreeding a universal threat?

His first statement of concern came three years after *The Origin*, as an afterword to his book *On the Various Contrivances by which British and Foreign Orchids are Fertilised by Insects, and on the Good*

Effects of Intercrossing. The last paragraph of that hefty work, most of it devoted to botanical minutiae, ends: 'Nature thus tells us, in the most emphatic manner, that she abhors perpetual self-fertilisation. This conclusion seems to be of high importance, and perhaps justifies the lengthy details given in this volume. For may we not further infer as probable, in accordance with the belief of the vast majority of the breeders of our domestic productions, that marriage between near relatives is likewise in some way injurious,—that some unknown great good is derived from the union of individuals which have been kept distinct for many generations?'

The idea that children born to related parents might suffer harm was already in the air. The first study of its risks came in 1851 when Sir William Wilde (father of Oscar) found, in work years ahead of its time, an increased incidence of deafness among the progeny of cousins. Sir Arthur Mitchell, the Deputy Commissioner in Lunacy for Scotland, had earlier claimed that in the inbred fishing communities of north-east Scotland the average hat size was six and seven-eighths, a quarter-inch less than that of their more open-minded agricultural neighbours; proof, he thought, of the malign effects of the marriage of kin upon the mental powers.

Sex within the household has a venerable history. The Pharaohs lived through generations of the habit in an attempt to preserve the bloodline of a God. Akhenaten, who lived around 1300 BC, first married his cousin Nefertiti, then a lesser wife, Kiya, and then three of his own daughters by Nefertiti and then (perhaps) his own mother. The story is confused by difficulties with sorting out quite who was who (and one of his supposed wives was in fact male), but incestuous affairs were without doubt common in ancient Egypt. Cleopatra herself may have been the scion of ten generations of brother–sister unions. The practice is condemned in Leviticus, where the Children of Israel were enjoined that 'After the doings of the land of Egypt, wherein ye dwelt, shall ye not do.'

The belief that the children of cousins are bound to be unfit, and the desire of all rulers to control their citizens' private lives, still fuels a jaundiced view of the joys of sex within the household. In 2008, a British government minister, in reference to the Pakistani population of Bradford, made the quite unjustified claim that 'If you have a child with your cousin the likelihood is there will be a genetic problem.' Many of his fellow citizens share that vague Galtonian sense that inbreeding is harmful. Most of their alarm rests on anecdote rather than on science.

All states are interested in how their subjects behave in the bedroom. For years, England based its marital rules on those of the Church of England, which descend from those of the Israelites, themselves established to put an end to the habits of the Pharaohs. In 1907, after hundreds of hours of parliamentary discussion, the statutes were at last clarified. The new legislation removed absurd anomalies such as the biologically senseless law that forbade a widower to marry his dead wife's sister but it also firmed up the prohibition against sex with close kin, be it father with daughter, or brother with sister.

Politicians often act on the basis of prejudice. Darwin did not. When faced with a scientific question – about sex or anything else – he set out not to speculate but to discover. To learn more about inbreeding he turned again to plants. *The Effects of Cross and Self-Fertilisation in the Vegetable Kingdom* appeared in 1876. It gives an account of experiments on a wide variety of hermaphrodite plants forced to mate with themselves. His verdict was clear: 'The first and most important of the conclusions which may be drawn from the observations given in this volume, is that cross-fertilisation is generally beneficial, and self-fertilisation injurious.' It was 'as unmistakably plain that innumerable flowers are adapted for cross-fertilisation, as that the teeth and talons of a carnivorous animal are adapted for catching prey'. The exchange of genes between unrelated individuals was the rule and selfing an expensive exception. What was true of plants must, he imagined, apply to animals, men and women included.

Flowering plants have long been known to have sexual habits more inventive than our own. Charles's grandfather Erasmus's poem *The Loves of the Plants* is a work of science in two hundred pages of Arcadian verse. Many lines deal with the balance between male and female interests ('Each wanton beauty, tricked in all her grace, Shakes the bright dew-drops from her blushing face; In gay undress displays her rival charms, And calls her wandering lovers to her arms' – in other words, this species needs a pollinator). His descendant asked deeper questions in plainer prose. He found that many plants live in a reproductive universe that would have shocked the shepherds of Arcady. They have a system of choice that transcends the familiar preferences of one gender for its opposite. Most retain their original nature as hermaphrodites, with male and female parts in the same flower, but they span the range from obligate self-fertilisers to others that make an absolute demand for pollen from another individual. Many among the latter group have to ensure that they do not accept genes from their close kin, and have imposed additional and refined laws of sexual choice upon their mates.

The cross-fertilisation book was a first step in the scientific study of sex. Fifteen years earlier its author had noted that 'We do not even in the least know the final cause of sexuality; why new beings should be produced by the union of the two sexual elements, instead of a process of parthenogenesis.' Not much has changed. We are still not certain how the habit persists in the face of its obvious drawbacks in terms of cost, stress and more – in Dr Johnson's famous words, its expense damnable, its position ridiculous and its pleasure fleeting. Why mate, when virgin birth ensures that your own genes have a guaranteed chance of survival? Parthenogenesis – virgin birth – guarantees that all the genes of those who indulge in it reach the next generation. It looks like the obvious solution but parthenogenesis remains rare – a few lizards and fish, and about one species of flowering plant in a thousand. For a hermaphrodite, a bout of sex with oneself also ensures that the DNA is not diluted with that of an unrelated individual. For creatures with separate males and females incest – sex with a close relative – is quite effective at keeping genes in the family, but that too is frowned upon.

Plants hint at the answer. Plenty among them have, like men and women, separate sexes. Some – like the strawberry or the dandelion – have gone to the opposite extreme for they are parthenogens and propagate themselves with shoots, roots or broken fragments. The majority of the flowering kinds have taken a lesser step towards asexuality for they are hermaphrodites that bear male and female functions on the same individual.

In spite of the chance they have for sex with themselves some hermaphrodites insist on exchanging genes with a stranger. Others in that situation are, in contrast, happy to self-fertilise and will accept genes from a different flower on the same individual, or evolve flowers with both male and female structures that can fertilise themselves. That pattern marks a real step towards the abandonment of sex.

Animals, too, have often tried to give up that habit. Some, like praying mantises and certain lizards, are true parthenogens, while a few such as snails and worms are hermaphrodites. Others go in for more bizarre forms of close copulation. The habits of mites would astonish any pornographer. A certain parasite of locusts gives birth to two types of male. The first clambers back into his mother and fertilises her. With their help she then produces a second brood, with a few males included – and those males then have sex with their sisters. In another mite, a mother has sex with her grandson, the scion of her own daughter, the daughter herself a child of the mother herself and a son. Other mites confine themselves to brother–sister pairs – but they copulate before they are born.

Darwin saw that, when it comes to sex, plants are more convenient subjects for experiment than animals. Self-pollination in hermaphrodites marks a biological indulgence more extreme than the matings between cousins that so concerned him, or between sons and mothers, fathers and daughters and brothers and sisters. He set out to explore how often it took place, what it did to the health of his subjects and, with luck, to learn a little about the importance of sexual reproduction in general.

Within a few months of starting work as a planned pollinator in the greenhouse at Downe he found that the effects of inbreeding could be dramatic. With the help of a botanical condom – a fine mesh to keep out insects – and a small paintbrush he could himself, like a bee, move male cells to the female parts of a flower and could arrange that the plant received its own genes, or those of another individual.

First, Darwin noted that certain species would self-fertilise, while others refused to do so even when obliged to try. Among those that did, he discovered – somewhat to his alarm – that the habit did damage later generations. His initial experiments were on toadflax, a common yellow-flowered weed. In the wild, out-crossing was the rule. In the greenhouse, he could force his subjects to self, and soon found a large, and unexpected, effect upon the next generation. The progeny of such crosses were smaller and less vigorous than were those of plants allowed to mate with another. At first he supposed that his inbred offspring were weakly because of some disease, or because they were grown in unsuitable soil. That was not so, for however well they were treated they stayed feeble. Darwin ran through a variety of species – carnations, tobacco, peas, monkey-flowers, morning glory, foxgloves and many other garden and wild flowers. With statistical help from Francis Galton he discovered that, almost without exception, those grown from crossed seed were taller, healthier and more productive than were those from self-fertilised. Some experiments went on for several generations, and the effects of sex with a relative got worse with time. The inbreds suffered most of all when life was hard: when they were crowded, had to compete with their outcrossed kin or were moved from the greenhouse to the rigours of the open air. The malign influence of selfing applied almost as much to species that went in for it in nature as to those that almost never did so.

Consistent as most of his results were, once or twice, out of thousands tested, the descendants of a self-fertilised individual were healthy and vigorous. Sometimes they even outgrew their competitors. A particular selfed morning glory (a species normally damaged by inbreeding) he referred to as 'Hero' because its line flourished under that mode of reproduction. Why, he did not understand.

Even so, in most plants in the wild, sex with a close relative appeared to be impossible, an expensive error or – at best – a doctrine of last resort. The implications for humans were, perhaps, alarming.

Darwin's views were, we now know, too inflexible. Almost all hermaphrodite plants have at least the potential to fertilise themselves and many do so as a matter of course (he also denied the importance of selfing in animals and was again mistaken: I myself once worked on hermaphrodite slugs, who manage quite well with sex within their own skins). Only about one in five hermaphrodite plants prefers to self as a general rule. Others that normally outcross will self when no alternative is available and only a minority avoids the habit altogether. The notion that grew up in the century or so after his book that plants could be divided into two distinct types based on sexual habits is wrong. In fact, they go from obligate selfers to determined outcrossers, but most are happy to adopt either practice as conditions change. Some shift from sexual to asexual in different places or as the seasons move on. For others selfing is a side-effect of sex, a kind of green onanism when a bee, as it flits from flower to flower, fertilises one flower with pollen from another on the same individual.

The decline in fitness of self-fertilised individuals and the occasional appearance of healthy lineages both emerge from the simple rules of inheritance (which were not known to the nineteenth century) and from the existence of large amounts of hidden genetic damage in most plants and animals. A hermaphrodite that bears two different versions of a particular gene – for example, a pea in which the DNA that codes for seed shape has the instructions for round seeds paired with another set for wrinkled – will, after self-fertilisation, produce half the next generation with the round–wrinkled mix, a quarter with round alone and a quarter with just wrinkled. If they are again selfed, the pure round and pure wrinkled plants will have offspring identical to themselves while those with both kinds of instruction will repeat the proportions that emerged in the previous generation – a quarter pure round, a quarter pure wrinkled and the remainder with a single copy of each variant. As selfing goes on, a smaller and smaller proportion of the population retains both versions of the shape gene. Generations of such crosses hence lead to the emergence of 'pure lines', within which every individual is identical: all with round seeds or all with wrinkled. The same is true for every other variant in each line. In time, each will contain different combinations of genes for seed shape, colour, height and so on.

To make a pure line is not easy, for if the original population contained hidden variants harmless in single dose but harmful in double (as many do), its descendants pay the price when such variants are exposed in double copy. The more damage is concealed, the smaller the chances of success. The effect can be spectacular. In loblolly pines, natives of the southern United States, just a fiftieth of eggs fertilised by pollen from the same plant survive. As a result it is almost impossible to make an inbred line.

Lines of genetically identical plants are at the centre of modern agriculture. Thousands have been produced for use as crops or as garden flowers. Wheat, rice, barley, tomatoes and more – all have their reproductive lives controlled by farmers and almost all are the descendants of a few survivors from vast numbers of inbred lines who paid a fatal price for the damaged genes hidden within their ancestors. The survivors were those who drew lucky in the sexual lottery. Darwin's discussions with breeders told him that the first few generations of kin mating caused the effects to get worse and worse – because, we now know, more and more harmful genes emerge in double copy as time goes on. Now and again, as in the famous morning glory 'Hero', an inbred line wins because it inherits, by chance, genes that increase, rather than reduce, its ability to survive. Its descendants thrive and may, in time, be used in their millions on farms or in gardens.

Farmers, consciously or not, have built on that observation and the same is true in the wild. In some species, well-adapted inbreds emerge to cope with the horrors of Nature. Millions of identical individuals then fill the landscape. The advantages of self-fertilisation often depend on how predictable life may be. The practice is commoner when the struggle to survive involves starvation or bad weather, which come back more or less unchanged each year. Perhaps, by chance, a set of genes emerges that deals well with food shortage or with cold. It then pays to stick to that well-adapted combination rather than to mix it with other genes during a spasm of sex. Selfing is more frequent in cold and starved northern forests and when life is short or when few mates are available. For hermaphrodites whose pollinators are grounded by bad weather, or if the season is so bad that it becomes hard to make a decent flower, a shift to selfing also makes sense, for the choice is between an unhealthy brood or none at all.

If – as in the tropics – the main enemies are parasites or predators, who themselves use sex to shift their own tactics by scrambling up their genes, that tactic does not work. For a plant to stick to a single strategy is to play poker with the same hand each time against an opponent who reshuffles on each deal. Sooner or later the other player will draw an unbeatable combination, and the selfer will go bust.

The rules that apply to plants work for animals too. When – in an echo of the Down House experiments – wild mice are mated brother with sister in the laboratory and the offspring released into nature almost none survive. Inbred animals do not often die of obvious genetic disease, but their parenthood can much weaken them. Song sparrows on a small island off Canada's west coast have been ringed for years and their pedigrees worked out in detail. Those born to close relatives are at more risk of death in bad weather than are other birds. The island of Soay, in the St Kilda group, is famous for its native sheep, which have been there since Viking times. The animals are filled with worms, and those with the heaviest burden suffer most of all when faced with a vicious Scottish winter. Lambs born to close relatives have more parasites, and are at more risk of death in storms, than others. Once purged with medicine, though, the inbred animals survive the tempest as well as do their outbred kin, as proof that the worms are to blame. The finches on the Galapagos also pay the price for sex with close relatives, as do wild shrews, red deer, seals, toads, bats and many other creatures.

Endangered species are, given the shortage of mates, at particular risk. Planned parenthood can help. The Florida panther was once on its last legs, with many animals plagued with kinky tails and undescended testicles because of the exposure of genetic damage in the tiny and inbred population that remained. In 1995, eight female Texas cougars (a related species) were introduced, and now the natives, with the help of their relatives' genes, have fought off their tail and testicle problems and returned to genetic health. The situation in zoos is just as bad. A mere hundred or so Arabian oryx were left in the wild by the 1960s, and most were soon killed by hunters. Two males and a female were rescued, and a dozen or so additional individuals were already in captivity. The entire world population of two thousand individuals, domestic and nominally wild, now descends from those few founders – and the most inbred individuals still have fewer young than average.

If sex within the family is bad for plants and animals, what might it do to people? Many are convinced of its dangers and many societies try to limit the practice. Every one of the United States has some restriction on marriage between relatives and cousin marriage is a criminal offence in eight states and in a further twenty-two is at least illegal, although the rules are often ignored. An attempt to ban such alliances was defeated in Maryland as recently as the year of the millennium. In a nod to the eugenic agenda, Wisconsin restricts them to couples in which the wife or husband is infertile, or where the wife is over fifty-five. The traditions of particular groups can lead to a reluctant acceptance of difference; uncle–niece marriages are allowed in Rhode Island but only for Jews, and male Native Americans in Colorado are allowed to marry their step-daughters even if elsewhere the habit is, in spite of the lack of any shared genes, against the law. Europe is just as confused. In Cyprus the Church prohibits marriage between second cousins but in Belgium incest is not even mentioned in the statute books. Britain, too, forbids incest and even prohibits the – rare – sexual liaisons between grand-father and granddaughter. Sweden, in contrast, permits half-sibs – children with one parent in common – to enter into marital bliss should they wish. The nation has even considered the legalisation of wedlock between brother and sister.

It is, needless to say, impossible to carry out planned crosses with men and women, but Charles Darwin came up with another way to test the dangers of inbreeding. First, he tried to get questions about cousin matrimony included in the 1871 census. He pointed out that 'the marriages of cousins are objected to from their supposed injurious consequences; but this belief rests on no direct evidence. It is therefore manifestly desirable that the belief should either be proved false, or should be confirmed, so that in this latter case the marriages of cousins might be discouraged.' His request was debated in the Commons but thrown out as 'the grossest cruelty', for it would cause children to be 'anatomised by science' (and from a parliamentary point of view the issue was almost traitorous, for the Queen herself had wed her cousin). A query about 'lunatics, imbeciles and idiots' was allowed but was dropped for the next census a decade later as most people refused to answer it. Darwin was annoyed by his failure to persuade Parliament to ask a scientific question and complained about 'ignorant members of our legislature'. His son George was even more caustic about 'the scornful laughter of the House, on the ground that the idle curiosity of philosophers was not to be satisfied'.

George Darwin set out to build on his father's work. From the records of *Burke's Landed Gentry* and the *Pall Mall Gazette*, together with a circular sent to lunatic asylums, he worked out the frequency of cousin marriage in various groups. Such unions were, it transpired, twice as common among noblemen as among the proletariat. His enquiry to the superintendents of asylums as to how many of those in their care were the scions of related parents was, as they pointed out, unlikely to pay off because of the mental state of their charges. Even so, George found no increase in the level of inbreeding among the patients compared with that of the general population (even if the Deputy Commissioner in Lunacy for Scotland did assure him that most of his nation's idiots were the children of relatives).

After his mixed success with lunatics, the young man went on to study the inmates of Oxford and Cambridge colleges. He chose the boat-race crews – 'a picked body of athletic men' – and asked how many had been born of cousins. After a correction for a falsified answer from the stroke of Corpus Christi College Cambridge, he found that there was indeed a slight shortage of such inbred individuals among top oarsmen compared with the general population. The same was true among sporting boys in the principal schools for the upper and middle classes. In both cases the numbers were small and the evidence not altogether persuasive.

Then George had an idea that might produce large amounts of information on the sex lives of the British people. He used surnames – an inherited character – to estimate the extent of marriages among relatives. Two people with the same name, particularly a rare title such as 'Darwin' or 'Wedgwood', are, he realised, more liable than average to descend from a common ancestor. Indeed, Sir William Wilde had already found that the parents of deaf children had a higher chance of a shared surname.

George Darwin himself did little with the idea, but surnames have now been analysed in their millions. They give a new insight into mating patterns. In most places, they pass, like the Y chromosome, down the male line. The fit between the two is real, but for several reasons not precise. Common names ('Jones' – which means 'son of John' – included) originate many times in different places. To confuse the issue further, children are adopted into new families, or agree to change their name for testamentary advantage. Some people take up a new tag because they do not like the label they were given and illegitimacy, too, is a problem. These frailties weaken the link between shared names and shared genes.

Even so, a random set of a hundred pairs of British men who each had the same surname showed a real tendency for them to have a common set of genetic variants on the Y. Two males who bore the same rare tag were far more liable to share a Y chromosome than were two sharing a frequent name. Names hence provide a real insight into genetic history – and the police are already interested in tracking down criminals by using DNA to search for surnames.

The failure with the census, and his son's ambiguous results from idiots and Etonians, suggested to Darwin that perhaps the effects of human inbreeding were less dire than he had feared. Honest as always, he admitted that 'my son George has endeavoured to discover by a statistical investigation whether the marriages of first cousins are at all injurious, although this is a degree of relationship which would not be objected to in our domestic animals; and . . . he has come to the conclusion that on the whole points to its being very small'. He removed his comment on its harmful effects from the second edition of the *Orchids* book.

The effects of cousin marriage on health were too small to be picked up by Darwin but the mass of information now available, from official records, from surnames and from patterns of shared genes, makes it clear that its influence cannot be ignored.

European aristocrats, like those of ancient Egypt, have long married their kin. In a world in which nobody mates with a relative, and with the births of parents and children separated by twenty to thirty years, each reader of these pages could have had, at the time of Charles Darwin's birth seven generations ago in 1809, a hundred and twenty-eight different ancestors – two multiplied by itself seven times. For almost everyone that figure is too high, for some marriage among those who have no idea that their spouse is a distant relative is inevitable. Even so, social pressure for matrimony within the household can do a lot to reduce that number. Alfonso, the Infante of Spain, who died in the 1960s, had just twelve – rather than more than a hundred – ancestors seven generations back, King Alfonso XII, a contemporary of Darwin, had sixteen, while plenty of others in that noble line had between fifteen and twenty great-great-great-great-grandparents – far fewer than expected in a sexually open society. The Spaniards are still keen on the pastime, with levels of cousin marriage well above the European average, and their nation has the most inbred villages on the continent, Asturias in the north being the most inward-looking place of all.

The habit also thrives elsewhere. Many of India's thousands of communities insist on alliances within their own group. The Dravidian Hindus of the south have encouraged sex between cousins, or between uncles and nieces, for two thousand years, and continue to do so, with one marriage in ten between a man and his brother's daughter. In large parts of Africa and the Middle East, a fifth of all alliances – and in some places even more – is between close relatives. The practice is also common in the world of Islam. The Prophet discouraged the idea, but he did marry off his daughter Fatima to her cousin Ali, which in the eyes of the faithful legitimises the habit. The tradition is not, contrary to popular belief, central to that religion but has become part of its culture. In Bombay, one Hindu marriage in fifteen is between close relatives, but for Muslims the figure is three times as high. Among the nomadic Qashqai people of Iran, the incidence of cousin marriage is three in four.

The picture in the West is quite different. Most of us rarely meet our relatives, let alone sleep with them. In most places less than one marriage in a hundred is between cousins. The English have long been among the least inbred nations in Europe. George Darwin had imagined that farm labourers 'would hold together very closely' but he was surprised to find how few weddings among kin there were in the rural districts of his time. England has never had a peasant class that sits on its own land for centuries and falls for the girl next door for lack of a better candidate. John Bull was a tenant. His master often forced him off and provided, as an unexpected bonus, a wider choice of sexual partners. In modern Britain, husbands and wives are, on the average, sixth cousins. Most share no more than a great-great-great-great-great-grandparent in common – someone who lived before Charles Darwin was born – and most have no idea at all that they are related. Only on islands, real or metaphorical, do we go in for the habit. Many Northern Irish travellers and a fifth of the people of parts of the Hebrides choose a first or second cousin as a mate. On Darwin's island today, faith is a better barrier to sex than are miles. In Bradford, the Pakistani community is among the most inbred in the world, with its British-born children having an even higher chance of marrying a cousin (often one still in Pakistan) than did their parents.

Social pressures can also reduce the extent of inbreeding. In many places the prospect of prison for sex with a cousin has, no doubt, often put paid to romance. In hunter-gatherer days, men and women roamed the landscape to about the same extent. With agriculture, life changed. The geography of men and women as seen in the patterns of Y-chromosome and mitochondrial genes shows that the males tended to stay at home while their mates came from elsewhere. The son inherits the capital and does not wish to move or share with a neighbour. He prefers a wife from far away who is, as an incidental, unlikely to share his genes.

Close inbreeding can – as Darwin had found in the greenhouse – impose a real burden. All human populations contain damaged genes, manifest only when inherited in double copy. The children of cousins may pay the price for that legacy. As Bagehot wrote: 'It has been said, not truly, but with an approximation to truth, that in 1802 every hereditary monarch was insane' and inbreeding was at least in part to blame. Every reader of these pages carries at least one gene in single dose that would kill them if present in double copy. Most inborn faults are hidden by normal versions of the same gene; for example, one British child in two thousand five hundred is born with two doses of the damaged gene that leads to cystic fibrosis, but one Briton in twenty-five has a single copy. If relatives mate and if their common ancestor bore a faulty piece of DNA, the chance that each partner will inherit that fault by virtue of shared descent goes up. Their children are then at increased risk of receiving two damaged versions.

The malign effects of sex within a closed pool were noticed long ago. Umar Ibn Al-Khattab, the second Caliph and direct follower of Muhammad, advised members of a certain tribe to marry out because, he believed, they had become weak and unhealthy through their habit of sex with kin. Its power became manifest in 1908 when Sir Archibald Garrod identified an inborn illness called alkaptonuria as a condition in which two copies of a faulty piece of DNA were needed to show their effects. An enzyme that breaks down certain food substances is damaged. Symptoms include dark ear wax, smelly urine and, later in life, heart, skeletal and other problems. The disease is rare, with just one case in every twenty thousand births, but Garrod found that more than half of the parents of his patients were cousins. The same is true for other such conditions. In France, with one marriage in five hundred between first cousins, the incidence of cystic fibrosis is over-represented by seven times in the children of such matings.

In an unfortunate coincidence, certain places with a lot of sex among kin – North Africa, the Middle East and the Indian sub-continent – also have a high incidence of inherited blood diseases that protect against malaria. Sickle-cell is carried in single dose by almost half the members of some African populations and related errors are almost as common in other places. Each is dangerous when inherited in double copy.

In Saudi Arabia, where in some villages eight out of ten people marry their cousins, such diseases are common. A fifth of all admissions to children's wards are due to hereditary disorders. Many families are unaware of the dangers and the devotion to cousin marriage remains. Doctors now advise those at risk to screen pregnancies in the hope that damaged foetuses might be picked up before they are born and the government insists that engaged couples are tested to see if both carry a blood disorder. No overt pressure is brought to bear, but the incidence of marriages within the family has dropped by a fifth.

Even in places without high levels of inborn disease children born to cousins die younger than usual. Such unions in Utah Mormons – not themselves an unusually inbred population – from the mid-nineteenth to the mid-twentieth century led to a notable increase in ill health. The effects became worse as the infants grew older, perhaps because death in old age has a stronger genetic component than do the accidents of infection or starvation that killed the pioneers' babies. Heart disease is also more frequent among the children of cousins. A study of half a million pregnancies in the modern United States suggests that the death rate of the sons and daughters of cousins rises by about 5 per cent above average. The products of uncle–niece marriages, a pattern frequent in India and elsewhere, do even worse and while incest – sex between sibs, or parents and children – is rare the offspring pay a high price. A German brother and sister, adopted at birth and strangers until they met as adults, had four children, two of whom were severely affected. A study of thirty or so Canadian children born to such parents also suggests that almost half inherit some abnormality.

A more subtle, but more marked, effect of within-family sex has emerged in Iceland. Among a hundred and fifty thousand couples born between 1800 and 1965, partners who were close relatives had more, rather than fewer, children than average. Even so, the proportion of the children of first and second cousins who them-selves reproduced (and hence the number of grandchildren born to the pair of relatives) was well below average, in part because many of those first-generation progeny died young. Charles Darwin and his cousin Emma may have been testimony to that effect, for seven of their ten sons and daughters expired before their time or lived on but stayed childless. Close mating may be more harmful to a family's prospects than was once supposed.

Continued inbreeding leads to a decrease in variability within a lineage. The effects soon extend across the entire genome. The overall level of DNA variation is lower in people who emerge from a limited pool of ancestors and, as a result, the level of inher-ited diversity gives an insight into the extent of inbreeding. Many illnesses – diabetes, heart disease and more – are more frequent, and more severe, among those so revealed to have a history of sex with kin. In Bradford, some members of the Pakistani community are uniform in long stretches of their genome. They pay the price in terms of health and even their general liability to infection goes up. In Darwin's day, childhood death came in the main from con-tagious disease. His beloved Annie died of tuberculosis (although the diagnosis was the obscure 'bilious fever with typhoid charac-ter') but her plight may, as he feared, in part have been due to her parents' marital history.

Sex is, needless to say, more complicated in the bedroom than in the greenhouse and the simple fit of health with kinship does not always hold. In some places, relatives marry to keep wealth within the household, which means that there must be some wealth around in the first place. In parts of India and Pakistan, cousin marriages are commoner among the affluent than among the poorest, for the latter have no financial incentive to set up home with a relative. The effects of cash outweigh those of genes. As a result, the children of cousins are less, rather than more, liable to suffer ill health. As is true for the sheep on Soay, the environment also plays a part. In the poor and embittered Japan of the 1940s, cousin marriage had a large influence on infant welfare, but twenty years later the effect had almost disappeared.

For both plants and people, sex (when not with oneself) must involve another party. Almost always, he or she must choose or be chosen from a pool of potential mates. The process calls for hard decisions. Some are obvious. Whites tend to marry whites and blacks blacks, while the rich marry the rich and the tall the tall and, some say, men tend to find a wife who looks rather like their mother. Plenty of religions make it hard to find a partner outside the creed. Language, place of birth, education and more also affect the choice of candidate. All this means that for any man or woman the number of possible mates is far smaller than it might appear to be. That observation is familiar enough, but biology sharpens up human sexual decisions in ways both obvious and less so.

The fundamental question about sex is: why bother? The habit is expensive for sex reduces the number of potential mates. It imposes the simple rule that only some individuals – those of a different gender from oneself – are available for copulation. Self-fertile plants circumvent that demand, but sexual creatures (ourselves included) have their choices much restricted by it. Often the rules become even more stringent. Evolution generates laws that ensure that no longer can any male mate with any female (or vice versa). They ensure, as a result, that an individual of one sex will be accepted by no more than a fraction of those of the other. Sex becomes more sexual than before. The process is driven by the need to avoid inbreeding.

Darwin discusses such issues in his book *The Different Forms of Flowers on Plants of the Same Species,* published a year after the volume on self-fertilisation. It begins with a simple tale: the story of the children of the village of Downe, who made necklaces from cowslips. They could, they told him, use just a few of the plants, those with a long 'pin' that protruded from the flower, through which they could thread the plants together. Other flowers, instead of a pin, had no more than a short protrusion called a 'thrum' and were of no use as juvenile jewellery. The cowslip's close relative, the primrose, was much the same.

Pins and thrums represent an additional mechanism of sex choice. Female 'pin' flowers were, Darwin found, much happier to accept pollen from male 'thrums' than they were from males of their own kind. The same applied in the opposite direction. A male needs a female, but the cowslip asks for more. The flower's form is inherited – which means that the plants decide whether or not to accept another's pollen advances on genetic grounds. Like mates only with unlike, so that a second sexual filter reduces the chance of an encounter between plants that bear similar genes. It is, as a result, a precaution against inbreeding. With pins and thrums, Nature has come up with a trick to reduce the proportion of individuals with whom genes can be shared. She has, in effect, invented more and more sexes.

Darwin's greenhouse experiments on cowslips have grown into a science that shows how, in both primroses and people, partners are chosen in unexpected ways and that the choice may decrease the prospects of successful mating between those who have recent ancestors in common. Thirty or so plant groups have evolved systems rather like that of the cowslip and primrose. He found others in which the flowers came in not two but three forms, each of which would not accept a partner from within their own class.

Now we know many more examples of such physical barriers to sex. Some species produce flowers that are mirror images of each other and can cross only left to right and right to left. In a bizarre system found in tropical gingers, some individuals are male in the morning and female in the afternoon while others prefer the opposite pattern. As a result they are obliged to exchange genes with those whose sexual timetable is different from their own. Once again, the imperative is to avoid carnal relations with those like themselves – with kin.

In his crossing experiments, Charles Darwin had no real idea of how and why his experimental subjects accepted or rejected particular kinds of pollen when he placed it on their female organs. He referred only to the 'extreme sensitiveness and delicate affinities of the reproductive system', which is poetic rather than persuasive. In fact, as in the cowslip and the tropical ginger, they make a test of kinship before deciding whether to accept a mate. The female parts judge the hopeful male cells by comparing their genes with their own. They reject any pollen grain if the similarity is too close.

The process is at work in the many hermaphrodites that insist on outcrossing. Like sperm, a grain of pollen contains but a single set of genes. Unlike the constituents of that potent liquid, the male sex cell from a flower must fight its way through a barrier of female tissue to reach the egg. To do so it grows a long pollen tube that penetrates into the appropriate part of its partner. Her protective layer bears the normal double complement of DNA. For her, to choose is simple: compare the pollen with her own tissues and if the two share too many genes, block it. For species that prefer to self-fertilise, the rule is relaxed or reversed.

For outcrossers, the system ensures that unrelated mates have the best chance of success. A new version of the identity cues carried by pollen is almost certain to succeed, for in its novelty it charms its way into the affections of all females, none of whom bear it themselves. As the generations go on, the new gene spreads – but it begins to lose its magic as more and more females inherit it and reject males with a matching copy. Each shift in male identity goes through the same process and in time a system emerges in which almost every individual has his or her own unique sexual calling card. That allows females to make decisions about the kinship of the hopeful males and to choose those most different from themselves. Other species delay that decision until later. The pollen is allowed to fertilise the egg, but matings with close kin are not allowed to develop further.

Animals – ourselves included – have a similar set of mechanisms, with a variety of genetic identity tests before sex is allowed. Some are obvious, while others are less so.

Simple familiarity can breed contempt. Unrelated Jewish infants brought up together in kibbutzim, or Asian children betrothed and made to live together when they are tiny, may prefer to avoid sexual contact when they grow up, and – in the latter case – are, after the arranged marriage, said to be less fertile and more liable to divorce than average. Brothers and sisters also tend not to fancy each other. Older sibs feel a stronger sense of aversion to their younger fellows than do the young to the old. The degree of kinship is the same, but the older child can be almost certain that the junior members of the household are the products of their own mother for they saw them cared for as babies. A younger sib, on the other hand, knows only that an older individual lives under the same roof – which could happen for other reasons. They are less repelled by the idea of sex with somebody who might not, after all, be a relative. It takes fifteen years of shared residence for a younger brother or sister to build up the erotic revulsion that an older member of the family can generate by watching a few months of childcare.

Social pressures play a large part in our marital patterns, but genes are involved too. Some are obvious – people do, after all, tend to marry someone of the same skin colour as themselves – but others are more subtle.

★

As a boy, I kept mice in my bedroom, a fad quashed because of the awful stench. At the time that was no more than a nuisance, but in fact the aroma of mouse urine was an introduction to a new world of sexual contact, through the nose. Quite unexpectedly, mice have more genes than we do. Almost all the extras are involved with the sense of smell. The genes that code for smell receptors – most of them decayed in the human race – are in full order. Mice have hundreds, which together can tell apart a vast diversity of scents. The animals choose both food and mates through the nasal passage.

Given the choice, an inbred laboratory female mouse prefers to mate with a male from a different line. So keen is she on a new swain that a pregnant female will resorb her foetuses to render herself available. Bedding soaked with male urine has the same effect. The females assess health as well as kinship. Their acute nostrils sniff out those who carry parasites and avoid them. Perhaps – as in the wormy sheep on the Isle of Soay – the healthiest males, with the most impressive statements of their fine condition, are less inbred.

Mice live in an aggressive sexual universe. Each male dominates a small patch in which he can monopolise the females, but their partners often hop over to a neighbour's territory for a change. The male marks his boundaries with urine and females base their choice on the same stuff. The more urine there is and the less familiar it smells, the better. The males are forced to engage in liquid battles in which each tries to water down the offerings of his competitors. The females go for the most productive and most aromatic among them. So potent is the identity cue that even the human nose, feeble as it might be, can separate some mouse inbred lines by scent alone.

The perfume is based on a series of proteins, coded for by related genes in four different families, one of which has over a thousand members. Two others are expressed in a special organ with its own set of nerves, at the base of the nose. Not only do the proteins have a strong scent of their own, but they bind other male pheromones to make a cocktail of desire. The genes involved are highly variable. As a result – just as in flowers – females can avoid males with the same odour, and the same family history, as their own. Like them, they steer clear of potential swains with low variability in the smell-related proteins, perhaps because their reduced and inbred state makes them less suitable as fathers.

Primates, too, signal with scent – which is why the aftershave industry does so well. Marmosets and tamarins, small New World monkeys, send out chemical messages with dozens of constituents to mark their territories, to advertise when they are available, and to bond with their partners. A male's brain lights up in response to female chemicals when she is most fertile. The largest response is in those parts of the marmoset brain associated in humans with emotion.

We smell, as any marathon runner soon finds out. Bloodhounds can sniff out individual identities and are confused by identical twins. Rats, too, can assess human kinship. The animals sniff for longer at an unfamiliar scent than at an odour that they have already experienced. Give them a sweat-soaked shirt and they can tell whether they have smelt it before. When tested with the scent of the brother of a familiar subject, they sniff less than when given a sample from a cousin. To rats, at least, we have an aromatic identity.

But can men and women, like rats, mice or marmosets, themselves identify the sweet smell of the opposite sex? The case is not proved. Many of the human genes for odour reception have rusted away, to leave fewer than half the number at work in mice, and we lack the special organ that is so sensitive to scent in other mammals (although a few of the genes that make it are still at work and will respond to mouse scents). Generations of students have sniffed T-shirts worn by women at different stages of the ovulatory cycle, with inconsistent results. In spite of the undoubted genetic differences that exist in the ability to taste certain chemicals it has been hard to obtain clear results on the role of scent in human mate choice.

Even so, some observations hint that – like dogs around lamp-posts – men and women do pass on romantic messages through the nose. Many perfumes contain synthetic musks of the kind used by monkeys or mice to choose a mate. One chemical, a relative of testosterone, has long been touted as a chemical messenger. The stuff is sold to farmers as 'Boar Taint' to test the sexual receptivity of sows. Some people can smell it while others deny that it has an odour of anything, but after several weeks of exposure even they begin to notice its presence and the number of relevant receptors in the nose goes up and up – and more in women than in men.

Mice, men and flowers have converged in their mutual distaste for sex with a relative, but how did such cues of identity evolve? There are intriguing similarities between the mechanisms of choosing a mate and those that fight off infection. An ancient tie between sex and disease may even be behind some of our own marital preferences.

All mammals, smelly or not, carry inherited identity cards on the surface of every cell. We cannot accept kidney transplants because our immune system compares the donor's genes with our own, recognises the tissue as foreign and rejects it. The less related the source of the organ, the fewer the genes in common and the lower the chance of success, which is why brothers and sisters are better donors than are pairs of strangers. The identity system is based on a set of genes that sit close together on the DNA. They live in a section that codes for the many functions of the immune system, our prime defence against infectious disease, and, as an incidental, against the novel challenges presented by tissue trans-plantation. Each comes in many different forms, which means that vast numbers of combinations are possible.

Disease is a potent agent of natural selection. Individuals with the most diverse set of immune-system genes, and those with large numbers of rare variants, tend to fight off infection better than others. Mice and even fish prefer to mate with those least similar to themselves in immune identity, as a hint that the tie between sexual choice and disease resistance is ancient indeed. In the fight against infection, such diversity pays, for the next generation will have, thanks to sex, a new mix of defensive genes, confusing the parasites' ability to evolve fast enough to evade our immune system. It hence pays to choose someone as different as oneself as possible.

As Darwin discovered, cowslips and other plants are very careful when deciding which pollen is acceptable, with a variety of devices to ensure that their reproductive parts stay free of cells from males physically similar to themselves. The erotic stink of mice does the same job and humans, too, may learn to avoid familiar kin. In truth, the sexual examination goes on well after the male cells arrive. Plants choose what pollen tubes are allowed to grow, and female insects may store the sperm of many males before deciding which should be allowed to travel further. Even after fertilisation, plants and mice are happy to abort a high proportion of their embryos, most of all those that arise from the attentions of a male relative.

The female reproductive system is a difficult and dangerous place for a sperm to find itself. Promiscuous mammals have longer vaginas than do those who stick to a few mates and make the male cells work harder to reach their goal. The vaginal tract is acid, too, and sperm do not much like that. In humans, of the millions implanted by a successful man, no more than a few hundred reach the neighbourhood of the egg, twenty or so make it to the point where they might be able to fertilise it and just a single cell gets in.

The smell of success lingers on after the sex act is over. Human sperm pick up and move towards chemical signals from the egg with the help of a gene that sits right inside the group that codes for smell perception. The complicated pore in the nose or the sperm cell membrane that picks up a single scent molecule, or a signal from the egg, each do more or less the same job and the two look remarkably alike – and, in a nod to their common heritage, some of the genes used by mice as they sniff the air to assess kinship by smell are also active in sperm. In an unexpected link between two sexual worlds, the sperm receptor also responds to the scent of lily of the valley and, given the choice, will swim towards it. Whether human eggs prefer to attract, or to allow entry to, sperm genetically different from themselves, we do not yet know.

Darwin's work on the sex lives of plants has strayed into fields that would have shocked his contemporaries. His interest in their reproductive habits grew from his concerns about the effects of inbreeding in humans and on his own family in particular. Its influence is real, albeit less severe than he had predicted, and both plants and humans have evolved mechanisms that limit its effects. Faced with the same set of challenges, natural selection has come up with similar solutions in both kingdoms of life, which would not have surprised him (although he would, perhaps, be startled to discover that human sperm are attracted by the scent of a flower).

The great man's concern about the possible damage done by cousin marriage to his own children was not justified. Of his sons, William became a banker and Leonard an army major. George was elected Professor of Astronomy and Francis Reader in Botany at Cambridge, while Horace set up as a scientific-instrument maker and was for a time mayor of that fair city. The naturalist's offspring married into several eminent clans including those of Keynes and Huxley and – in spite of their progenitor's concerns about inherited feebleness – have produced dozens of descendants eminent in science, medicine and the professions. They stand as living proof that intellectual aristocrats, unlike their botanical and blue-blooded equivalents, need not pay the price of keeping their biological heritage in the family.

CHAPTER V

THE DOMESTIC APE

'Let them eat cake!' said the Queen, and they did. Two centuries after the demise of Marie Antoinette, the poor are fat and the rich thin. Across the globe death from excess has, for the first time in history, overtaken that from deficiency. Eight hundred million people are hungry while a billion are overweight. The problem comes from evolution, as manipulated by man.

Darwin saw how farmers had bred from the best to produce new forms of life and used that notion to introduce the idea of natural selection. His argument is set out in the first chapter of *The Origin of Species*. Given time, and with conscious or unconscious selection of the best by breeders, new and modified versions of creatures from pigs to pigeons will soon emerge. Were they to be found in nature rather than in sties or lofts many would be recognised by naturalists as distinct species of their own.

In *The Variation of Animals and Plants under Domestication*, pub-

lished in 1868, Darwin went further in exploring the tame as the
key to the wild. The book speaks of ancient times, when 'a wild
and unusually good variety of a native plant might attract the
attention of some wise old savage; and he would transplant it, or
sow its seed'. That interesting event – the choice of favoured parents to form the next generation – was a microcosm of what had
moulded life since it began. The variety of breeds seen on the
farm was, he wrote, 'an experiment on a gigantic scale', both a
test of his theory and a proof of its power.

Savages have been replaced by scientists. Their work has produced many new varieties of plants and animals and, on the way,
has revealed the eccentric history of the food on our plates.
Modern biology has transformed farming. Planned breeding –
directed evolution – has led to an enormous drop in the effort
needed to feed ourselves. The British spend a sixth of their
income on breakfast, lunch and dinner, a proportion down by half
in the past five decades and by far more in the past five centuries.
For most people, shortage has given way to glut and for many citizens of the developed world food is in effect free.

The blessings so brought are equivocal. The real price of sugar,
starch and fat – high energy but low-quality comestibles – has
plummeted. Famine disguised as feast has spread across the globe.
Evolution on the farm transformed society ten millennia ago and
is doing the same today. Farmers have been powerful agents of
selection on wheat, maize, cows, pigs, chickens and more, but the
influence of those domestic creatures on the biology of the farmers themselves has been almost as great. Diet began to act as an
agent of natural selection as soon as the wild was domesticated ten
thousand years ago and caused people to evolve the ability to deal
with new kinds of food. Today's shift in what we eat will have
equally powerful effects on the genes of our descendants.

A new global power – and a new agent of natural selection – is
on the move. The empire of obesity began to flex its stomach in
the 1980s and shows no sign of retreat. Twenty years before that

dubious decade there was, in spite of a collapse in the real price of food, little sign of the coming wave of lard. Then, thanks to technology, came the industrialisation of diet; the last step in the scientific exploitation of the Darwinian machine. Now, a tsunami of fat has struck the world and its inhabitants are paying the price.

It does not take much to alter a nation's waistline. The rise in American obesity over the past thirty years can be blamed on an increase in calories equivalent to no more than an extra bottle of fizzy drink for each person each day. At the present rate two-thirds of Americans and half of all Britons will be overweight by 2025 and Britain will be the fattest nation in Europe. Among industrial powers, only China and its neighbours are insulated from the scourge.

The twenty-first-century plague is a side-effect of the triumph of scientific agriculture. Many of those worst afflicted suffer because they bear genes that make it hard for them to deal with the new diet. Many of the obese will die young or fail to find a mate. As a result obesity will soon be – as farming itself was when it began – a potent cause of evolutionary change.

The people who laid out the first fields lived above the rivers that snaked across a green and leafy Levant. For millennia they hunted game and gathered seeds as man had done for the whole of history. Just after the peak of the last ice age the Middle Eastern weather became wetter and warmer and the grasses flourished. The gatherers prospered. Thirteen thousand years ago came a nasty shock, for the climate turned cold and dry for several centuries. The chill persuaded people to plant grains, rather than just to collect them. Soon the thermometer went up once more, the crops flourished and agriculture made its presence felt. Within a few centuries, the Fertile Crescent was filled with tillers of the soil.

A similar way of life, based on maize and rice rather than on wheat and chickpeas, soon got under way in South America and China and, in time, even in Papua New Guinea, where banana and sugar-cane cultivation emerged six and a half thousand years ago. The habit spread fast. Farming reached Britain some four thousand years ago. The shift to the new economy was quite rapid, and the pursuit of wild game was more or less replaced by agriculture within just a couple of centuries, although people still ate plenty of seafood (and that remnant of the chase persists today). As new crops emerged the locals began to husband animals that could feed on them. Soon a hundred people could live on the space that had previously supported but one.

The new economic system led to a grand simplification of diet. *Homo sapiens* has eaten some eighty thousand kinds of food since he first appeared on Earth. A dig in Syria of the homes of hunters who lived just before the new economy emerged revealed a hundred and fifty varieties of edible fruit, grain and leaf in that single society. Even in the nineteenth century, Queensland aborigines feasted on two hundred and forty different kinds of plant. As the new way of life spread, the cuisine became simpler. Within a few years, the Middle East had just eight crops: emmer and einkorn (antecedents of wheat), barley, peas, lentils, bitter vetch and chickpeas. Quite soon the people of the whole world considered together ate no more than half the number of plants once used by a single hunter-gatherer band. In most places just a couple of crops – rice, maize or wheat included – became the staple food. They kept that status for ten millennia.

Now, things have changed once more. Some lucky citizens have taken a great leap backwards. The middle classes have returned to the hunter-gatherer diet. They forage in priccy supermarkets for an eclectic range of edibles, from avocado to zucchini, imported from across the globe. The revolution of the rich began soon after Columbus, when exotic delicacies such as potatoes, peanuts and tomatoes were brought from the New World. Other delicacies went the other way, albeit sometimes after a long delay; broccoli, for example, was almost unknown in the United States until the

1920s. On both sides of the Atlantic, those who can afford it have put ten thousand years of dietary history into reverse.

The advocates of avocado are still in a minority. Many of their fellow Americans or Europeans have meals almost as dull as those of the first peasants, without the privilege of growing the raw materials themselves. Just as at the dawn of agriculture, their choices are narrower than were those of their parents and grandparents. Cheeseburgers, chips and sweet drinks are full of cheap energy and the poor have seized upon them. Nowadays, the British obtain twice as many calories from fats as did their immediate forebears and on average the intake of sodium has gone up by ten times and that of calcium down by half compared with earlier times.

The junk food revolution tells the tale of artificial Darwinism in all its details. The taming of the hamburger also shows how man, the most domestic creature of all, has paid a high price as he tests the biological limits set by his own evolution.

The first farmers, like the modern poor, became less healthy as their dietary options shrank. The symptoms were different from those of today, but the causes – an abundant but inadequate cuisine – were the same. Their bones show signs of deficiency disease and the average height of adults dropped by fifteen centimetres as the new way of life spread. The loss was not regained for several thousand years. In North America, where maize became the basis of almost every meal and where it was worshipped as a god, another problem was a shortage of iron, for maize lacks the mineral and also interferes with the ability to absorb it from meat. Many people became anaemic. No doubt they were tired, weak and depressed as they pursued their wretched lives as tillers of the soil. Deficiency and its diseases – lassitude, infirmity and sadness included – have returned, but disguised as excess.

Thirty thousand premature deaths a year in the United Kingdom are due to an expanded waistline and ten times that number in the United States, where, in 2005, obesity overtook smoking as the main preventable cause of mortality. It is more

than a coincidence that as America's spending on food as a proportion of national income went down by almost half, that on
health care was multiplied by three times. In Central and Eastern
Europe, even more healthy years of life are lost per head than in
Britain. The present generation of men and women – those who
grew up before the new age of edible trash – may be the longest-
lived in history.

The problem for their children is fat. Medicine has long
known how dangerous the blight can be; in Hippocrates' words:
'Corpulence is not only a disease itself, but the harbinger of others.'
Thousands have died before their time of heart disease, stroke,
cancer and diabetes, the four horsemen of the obese. Many others
suffer from gout, arthritis, bladder problems, reduced fertility and
the other conditions that affect the fat far more than the thin.
The most dangerous effect of gluttony is to grow to resemble an
apple rather than a pear, for extra inches on the waist are much
more harmful than the same on the backside – and the apples are
taking over from the pears even among women, who used to
put more on the bottom than the belly as their weight went up.
The apple brigade store fat around their livers, where it is readier to
spring into action and to release fat itself, hormones and agents
of inflammation into the bloodstream.

In the modern United States, as in the New World at the dawn
of agriculture, Native Americans have paid a particularly high
price for the change in diet. A century ago, many kept to their
traditional cuisine. The Pima Indians of Arizona – the Corn
People as they called themselves – were thin as they ate their
hearty meals of tortillas or porridge, based on maize. Now, they
gorge on burgers, chicken and sweet drinks instead. In some ways,
however, their food input has not changed for the Pima eat just as
much maize as did their grandparents. The difference is that today
it has been through a cow, a chicken or a soft-drinks factory first.

Cheap corn gave birth to fast food. One American meal in five
is eaten in the car and the maize needed to feed its four passengers

with a cheeseburger each would more than fill its tank. A Chicken McNugget has thirty-eight ingredients – and thirteen come from maize. The fizzy beverage that washes it down is based on corn syrup and the raw material of the post-prandial milkshake comes from a cow fed in a yard, on maize, rather than in a field, on grass. The 'natural strawberry flavour' that adds its dubious tang is natural only because it is made from corn and not synthesised from chemicals. A quarter of the food items in American supermarkets now contain maize, and their rows of cheap packaged products – and thousands are introduced each year – bear witness to the second agricultural revolution that has taken place in the lifetime of most readers of these pages.

Seed crops – maize most of all – transform sunshine into food. Even better, they are easy to store and to move. Cows evolved to eat grass in fields, but now it makes more economic sense to feed them on grain on giant lots. More than half the maize and soy grown in the world is eaten by animals. As a result, global meat production has gone up four times since the 1960s, and the amount of flesh available per head has doubled.

Scientific farmers have done in a few decades what took peasants centuries to achieve with no science at all, but the early farmers' approach was, in its essence, identical to that of modern technologists. They understood little of what they were up to and may not even have made the tie between sex and reproduction. By the Middle Ages, the idea that attributes ran in families was accepted; as the 1566 book *The Fower Cheifyst Offices Belongyng to Horsemanshippe* put it: 'it is naturally geven to every beast for the moste parte to engender hys lyke'. Soon artificial selection, conscious breeding from the best, was under way (even if the horse-racing and dog-fancying fraternities clung to the odd idea that qualities were passed only down the male line). In the eighteenth century, English improvers became aware of the need to mate animals of equal 'beauty' and agricultural science was born. Robert

Bakewell, chief among the breeders, was frank about his motives. He called his barrel-chested New Leicester sheep 'machines for turning herbage . . . into money' and hired out his rams for stud at £1000 a season – a huge sum for those days.

Now animal breeding has become a massive business. Champion bulls and stallions can sire thousands of offspring, and new statistical methods allow their young to be compared over hundreds of farms to see which have done best. Often, the actual genes that lie behind their talents are not known. Milk yield in cows has doubled since the 1940s, but the sections of DNA that did the job stayed hidden for sixty years. Molecular biology is beginning to change that, with the DNA sequence of most domestic animals now complete, together with maps of hidden diversity that can track down where the most productive variants might be. The annual gain in meat or milk production brought by genetics is, in the developed world, around 1.5 per cent a year, well over a billion pounds' worth in Europe alone. Artificial aids – mechanical cows into which bulls can ejaculate and have their semen smeared across the globe, cloned sheep, engineered crops and more – promise wondrous things. Even so, with the consumption of meat expected to double in the next decade that will not keep up with demand. Plant technology has been even more successful and many genes for high yield or disease resistance have been tracked down, with many brought in from the wild relatives of our domestic species. Agriculture now works with foresight, a talent quite unknown to evolution but used, at least subconsciously, by the first farmers of all.

What did it take to become domestic? The basic demand is for a creature able to coexist with man. Men can choose, often without much thought, the most favoured individuals to found the next generation. For both plants and animals, improvement becomes inevitable.

Darwin knew little about the origin of fruits, grains and vegetables: 'Botanists have generally neglected cultivated varieties, as beneath their notice. In several cases the wild prototype is unknown or doubtfully known; . . . Not a few botanists believe that several of our anciently cultivated plants have become so profoundly modified that it is not possible now to recognise their aboriginal parent-forms.' Now the aboriginals have been found, hidden in their descendants' DNA.

All crops have a lot in common. From tomatoes to barley and from chickpeas to plums, the domestics are less diverse than their wild predecessors, grow taller and less branched and have fewer but larger fruits or grains that taste less bitter than before. They flower at different times of year and their seeds spring into life at once when planted rather than (as do those of many of their wild fellows) demanding a long rest.

The tale of all those changes is hidden in the DNA. The story of maize – the raw material of junk food – shows how biology can reveal the past. Maize descends from a wild plant moulded not long ago into a dietary staple so different in appearance from its ancestor that for many years its origin was unknown.

Darwin himself knew that maize was ancient, for on the *Beagle* voyage he found cobs embedded in a beach raised by slow upheaval many metres above the sea. Its story began in southern Mexico around eight thousand years ago when people began to harvest, and then to grow, a wild grass called teosinte, the 'grain of the gods'. The tale of maize is that of the New World. Teosinte is still abundant over large parts of South America, even if several of its dozen or so species are under threat. Male and female organs are held in different places on the same individual, with a 'tassel' that bears pollen, and a number of small spikes that carry the female parts.

At first sight, the wild version looks quite unlike the familiar corn on the cob of today and was once assumed to be a relative of rice instead. A teosinte cob – a family of seeds held together on the same structure – is little more than twenty-five millimetres long, compared with thirty centimetres or more for its cultivated equivalent. When mature, the teosinte seeds, each within its own hard coat, form an 'ear' with half a dozen or so separate segments. The coat protects them from the digestive juices of the animals and birds that eat the cob. Each seed breaks off when ripe and, with luck, passes through the gut, falls on fertile ground and germinates to form the next generation.

The maize cob, in contrast, has five hundred or more kernels. The seeds are larger than before, come in a variety of colours and contain far more starch. They do not fall off without help and lack a protective outer sheath. They are, as a result, digested, rather then excreted, should they be eaten. If, at the end of the season, the whole cob is not harvested but falls to the ground, it bears so many seeds that almost none survive the intense competition for light and food. Maize is, as a result, entirely dependent on its human masters for reproduction. The plant has changed to such a degree that it looks quite unlike its ancestor.

Even so, the kinship of maize and teosinte is still close enough to allow certain wild strains to hybridise with their tamed descendants (farmers hate the idea for it degrades their crop, but scientists use it to rescue valuable genes before the natives disappear). The DNA of the modern crop is closest to that of the teosinte that grows in the hills around the Balsas river basin in south-west Mexico. There, McDonald's finds its roots. The oldest known cobs, six thousand three hundred years old, come from a cave in the valley of Oaxaca, four hundred kilometres away. At about that time, the people of South America began to thrive on their tamed grass. They soon learned to treat it with lime to release its essential vitamins – a talent forgotten until the mid-twentieth century, when the deficiency disease pellagra was tracked down to a diet of untreated maize.

At least a thousand genes in modern maize differ from those of teosinte. Fossil DNA from seeds four and a half thousand years old shows that, even by then, the farmers had already selected genes to improve grain quality and size. Just five genes, or groups of genes, were responsible for most of the shift towards the domestic. Many more play a smaller part. The move from grass to food

involved mutations that change the slim side-branches of the grass
into stout maize ears, others that remove the hard case around
each seed, while yet others ensure that the grains stick to the cob
and do not shatter when touched. Long stretches of DNA on
either side of those points scarcely vary at all, as a hint that large
blocks of inherited material were dragged through the population
by breeders as soon as the new attribute was noticed.

Maize improvement has become an industry. The plant is the
most widely cultivated crop in the world, with three hundred million tons grown each year in the United States alone. It has been
mutated, selected and hybridised to give hundreds of distinct
strains. Some are tall – seven metres high – and some short, some
large, coarse and used as cattle fodder, with others selected to have
tiny ears, the size of those of teosinte itself, and just right for a
cocktail snack. Sweet-corn is full of sugar. The starch itself, in
some kinds, bursts apart when heated, to give popcorn. The plant
now flourishes from the far north to the tropics and is far more
productive than its ancestors of even fifty years ago. The science
of maize has changed the global economy as much, or more, than
has nuclear power.

The maize genome has a bizarre and unexpected structure. It
contains almost as much DNA as our own and can boast of
twice as many genes. Most consists of bits of mobile DNA that
invaded long ago. Some of those molecular parasites can no
longer copy themselves and sit sullenly in place, while others
wake up now and again and move to a new site. They can cause
mutations as they go or capture a functional gene, altering its
effects as they do. Many of the mutations involved in the
improvement of maize emerged from this constant flux. Maize
DNA still changes fast. Some inbred lines descend from a shared
ancestor that lived just a few decades ago, but are already as distinct from each other as are humans and chimpanzees. The
mobile elements have been so active that, when two inbred lines
are compared, on average a fifth of all genes differ in where they

sit on the chromosome. Maize, plain food as it is, has a complicated biology.

Other crops have a less chequered history. Apples are easy. Fifty
years ago, Almaty, in Kazakhstan, was – like Norwich in Tudor
times – 'either a city in an orchard or an orchard in a city'. Its
name means 'father of apples', but the place is now more
notable for its Porsche and Mercedes dealerships. The city and
its surrounds were the site of a vast domestication. The genes of
the chloroplast – the green structure found in leaves – show that
the apples we eat today are almost all the descendants of just two
ancient Kazakh trees. Those mothers of all the world's apples
grew not far from Almaty. Wild trees, some as big as an oak, are
still scattered through the Tien Shan Mountains nearby. They
are part of what was once a vast fruit forest, the home of the
snow leopard, filled with walnuts, grapes and apricots as well as
apples. Today's varieties, from the insipid Golden Delicious to
rare strains such as Zuccalmaglio, have emerged through mutations and selective breeding in the lines that trace their ancestry
from those two progenitors. They are maintained with grafts
and cuttings.

Unlike the small and bitter crabs borne by most wild apples, the
fruits of the Tien Shan are large and sweet. They became luscious
when the trees changed their reproductive partners. The seeds
of most wild apples are moved by birds that peck at the fruit,
but in the Tien Shan, the Mountains of Heaven, bears do the
job instead. Both animals eat fruit and both scatter seeds in
their excrement. A bird is happy with a small reward but a hefty
mammal demands a more substantial bait. The trees grew sweet
apples to oblige. Eight thousand years ago, people and their horses
moved into the fruit forest and developed a taste for the ursine
delicacy. Kazakh apple seeds travelled in horse and human guts
down the silk roads that skirt the mountains. Now, their descendants fill supermarkets across the globe. The peach also traces its

origin to a wild mountain landscape in western China and reached Europe only in Greek times. It has diverged, like many other fruits with stones, into a variety of forms since then.

The potato has a more restricted history. It finds its home in a small patch of land, in Peru, north of Lake Titicaca. It has been cultivated for five thousand years and has diverged into a large variety of forms (to underline its importance, the United Nations Food and Agriculture Organisation defined 2008 as the International Year of the Potato). Lentils and chickpeas, too, each descend from just a single wild ancestor, as do peas. The various strains of rice, in contrast, emerged from two or three distinct species of wild grass cultivated in China. Wheat is different again, for the modern crop emerged as a result of crosses between several species of grass, some still around today, which came together to generate a plant with many more chromosomes than before.

Not long after wheat, chickpeas and the rest appeared on the plate, wild beasts were invited into the household. No more than a few accepted the offer and most of them had done so before the time of Christ. Quite soon society was transformed. Cows, pigs, horses and sheep became every farmer's treasured possessions, and a lot of effort was devoted to keeping them happy. As soon as they abandoned the wild, the animals began to change, and all in more or less the same way.

Farm animals, of whatever kind, tend – like their botanical equivalents – to follow some general rules. They are smaller and more lightly built than their unbroken counterparts, with shorter faces and smaller jaws. Often, they vary more in colour and shape than before, and many develop spotted coats. They are fatter, with longer intestines, breed through the year and make more milk. Most show less of a difference between males and females than in the wild. In many of their ancestors – wild cattle, horses and pigs – a large part of the force of natural selection involves differences in sexual success. Battles among males lead to the evolution of expensive horns or tusks, with days spent locked in combat. Once sex is under human control that wasteful effort can be directed to the production of milk, meat or wool instead, which is why domestic bulls or rams are less infuriated by their rivals than are their untamed relatives. In their lazy lives they tend towards promiscuity rather than the faithful bonds some of their ancestors preferred – which is useful for farmers when they wish to choose particular animals as parents.

One species in particular was quick to abandon its ancestral habits. It was the first to accept servitude and has used its own personality to manipulate mankind. It reveals, more than any other, quite what it takes to become tame.

Darwin was a dog-lover. He devotes the first chapter of *The Variation of Animals and Plants under Domestication* to the history of those creatures. So great was their diversity even in his day that he was uncertain whether dogs had descended (as in fact they do) from a solitary ancestor, the wolf, or from several, with the fox and jackal as additional candidates (although he did dismiss the widespread view that each breed had descended from a separate wild ancestor, now extinct). As he points out in *Variation under Domestication*, even barbarians attend to the qualities of their pets, to such a degree that the dogs of Tierra del Fuego have gained the instinctive ability to knock limpets off rocks. The breeders were often ruthless: the book tells of Lord Rivers, who, when asked why he always had first-rate greyhounds, answered, 'I breed many, and hang many.'

As Darwin noted, the dog is now the most varied of all mammals, both in mind and body. Some breeds were ancient. On a visit to the British Museum Darwin identified images of a Mastiff on Assyrian monuments from the sixth century BC. Others were more recent and had diverged much more from their wild ancestor. Some attributes – such as the shape of the head and the receding jaw of the Bulldog and Pug – might, he suggested, have arisen as sudden 'monstrosities' (or mutations, as we would call them), but the majority came from the slow accumulation of favoured forms. The dog was a marvellous model of how flesh can be moulded by human choice.

DNA shows that all dogs are the descendants of wolves, which will still cross with them when they get the chance (*Domestication* tells of 'the manner in which Fochabers, in Scotland, was stocked with a multitude of curs of a most wolfish aspect, from a single hybrid-wolf brought into that district'). The earliest bones found with those of humans are in a German dig some fifteen thousand years old, and the animals probably loitered around camp-fires long before that – which means that they entered the household well before any other creature. Even in their first days they changed, with shorter legs than their vulpine ancestors as a hint that they no longer roamed the countryside.

Since then, the animals have been subdivided into a wide variety of forms. Four hundred breeds are recognised and a hundred and fifty have official pedigree societies. Such organisations keep a close eye on their pets' sex lives and their rules often insist that both parents must belong to a rigidly defined type and that any dubious bloodline must be thrown out. Such exclusivity can lead to rapid change.

Certain breeds such as Mastiffs, Chows and Salukis have been distinct for centuries (even if the Pharaoh Hound, with its pointed ears and short coat that resemble those of the images on Egyptian tombs, is in fact a fake; a modern copy of an extinct breed). Most, however, are less than four hundred years old, and many are even younger. Their vast diversity is witness to what human choice can do to a once-wild animal.

In 1815, there were no more than about fifteen designated dog breeds in Britain. The first formal dog show was held in 1859, the year of *The Origin*. By then the numbers of breeds had risen to around fifty. Many of the most popular of today's hundreds of varieties – terriers, spaniels, retrievers and so on – trace their origin as distinct breeds no further than the past century, which means that they have gained an identity in no more than fifty or so canine generations. The genes show that almost all were founded by fewer males than females, evidence that – in the ancient tradition that quality passes only through fathers – a popular sire was mated with many bitches. Some males still have over a hundred litters, a pattern at variance with the monogamous sex life of the wolf. The breeders hold to their eccentric belief in the power of sperm over egg and by choosing only the very best as fathers much reduce the size of the available population.

Sometimes, a single mutation can spark off a new variety. The largest dog, the Irish Wolfhound, stands a metre high at the shoulder and tips the scales at well over fifty kilograms. Sixty Chihuahuas would fit into a single Wolfhound – but the difference in size is due to a single gene, which comes in one form in the big animal and another in the small. The Whippet is a racing dog. It too owes some of its identity to a simple genetic change. Now and again a heavily muscled individual – a 'Bully Whippet' – appears in a litter. It bears two copies of an altered gene for a muscle protein and is much misshapen. Most such pups are killed at birth. Many other members of the breed carry just a single copy of that gene. They are faster than average and the gene was unwittingly selected for as the animals were bred for speed (and it has now revealed itself in beef cattle and even in a young German, whose mother was a champion sprinter). Perhaps the most repellent of dogs is the Mexican Hairless, or Xoloitzcuintli, first bred for food and also used as a bed-warmer by the Aztecs. As its name suggests, the creature is entirely bald. A mutation in a gene which in humans leads to loss of hair and sweat glands is responsible (and, unusually enough, ancient statues show that the error has been around for three thousand years). Darwin himself identified a family with the condition – and almost worked out the pattern of inheritance, for he noted that it was passed through daughters but expressed only in sons, which is exactly what is expected from its position on the X chromosome. It was, many years later, the first human gene to be precisely located on that structure.

A few other dogs, Dachshunds included, also owe their identity to simple inherited errors. For most of the named forms, in contrast, divergence involves many genes. As they build up each type gains its unique appearance. Canine diversity is arranged in a way quite distinct from our own. People, wherever they come from, are more notable for similarities than for differences, but a large part of the variability among dogs as a whole emerges from divergence among breeds. The pedigree clubs have, the double helix proves, been real barriers to the movement of DNA. That in turn has led to intense inbreeding within particular lines. The three hundred thousand Golden Retrievers in Britain trace their descent in the past thirty years to no more than seven thousand males. Other kinds have lost nine-tenths of their total variation in just the dozen or so generations since the 1970s. The dogs have paid a high price. Determined – or deranged – insistence on forcing each line towards an arbitrary standard has led to King Charles Spaniels whose brains are too big for their skulls and Pugs whose eyes pop out so far that they are scratched whenever the animal bumps into something. Pugs are so inbred that the ten thousand in Britain share recent ancestry with only around fifty animals.

In spite of the genetic damage they suffer, dogs have exploited humans very effectively. They do not pay with their lives, or the products of their bodies, for food and shelter. No other creature is so tied to its master and no other domestic animal has been so subdivided. Most farm animals joined the family far later than did the dog, and some walked into the fields on several occasions and in different places.

A hamburger has a chequered history. As the *Domestication* book notes, the cow was tamed on two continents, in Africa and in the Middle East. Cattle were precious long before they were farmed. In the caves around Lascaux, in southern France, are images of more than a hundred aurochs, its gigantic ancestor. That impressive beast roamed wild in Europe, North Africa and parts of Asia and lasted until the 1620s. Sumerians had a cow goddess, the 'Midwife of the Land and Mother of the Gods'. Later came the Semitic deity Ishtar, whose bull–god partner made enough semen to fill the Tigris. Egypt, too, had a bovine obsession. The Pharaoh was called the Mighty Bull, with a tail on his kilt and bull legs on his throne. The Israelites for a time worshipped a golden calf, and suffered divine displeasure for their ways. The cults of the Minotaur, the toreador and the Western show how the animal retains its emotional power – and some golfers, it is said, still use a desiccated bull penis as a lucky putter.

The bones of domesticated cattle appeared in the Middle East around nine thousand years ago and in Europe around 5500 BC. Cows continued to mate with wild bulls for thousands of years. Ancient DNA shows that the female, mitochondrial, lineages of today's European cattle are quite distinct from those of their aurochs ancestors, while their Y chromosomes, the indicators of male ancestry, resemble the male chromosomes of the huge and extinct bovine. The wild bulls must have continued to impose their desires upon the domestic cows, with or without man's consent.

Bacon sandwiches tell a different tale. Pigs came into the household on several occasions, in different parts of Europe and the Near East, with some later input from Asia. Fossil DNA reveals a wave of Near Eastern pigs that moved into Europe and was then replaced by a taming of European wild boar.

Horses, too, have several origins, one of which is close to the home of the apples in Kazakhstan. Traces of mare's milk (still popular in that country) on five-and-a-half-thousand-year-old pottery fragments suggest that they were tamed by then. The sex bias by the horse-breeders has been extreme. Ninety-five per cent of the three hundred thousand racehorses alive today bear the same Y chromosome as evidence of descent from a single stallion. His name was the Darley Arabian, who was brought to England from Syria in 1704 by the then British Consul, Thomas Darley (two others, the Byerley Turk and the Godolphin Barb, provided almost all the other chromosomes). Europe has no native sheep or goats and the domestic forms had a simple and single origin in the Middle East.

The chicken, whose rendered flesh is a staple of the junk-food diet, descends from two or more species of Asian jungle fowl – and is now, with a population of almost thirty billion, the commonest bird in the world. It spread from its home in Thailand to fill Europe and the far Pacific and from there to reach South America before Columbus. In spite of their long years on the farm the birds retain a lot of diversity, perhaps because, like cattle, they continued to cross with their wild relatives. Their reproductive lives have altered more than those of any other creature, for some lay ten times as many eggs a year as do jungle fowl, each egg twice as heavy as those of their ancestors.

All these animals changed their minds as much, or more, as their bodies as they left the wild. Zoo animals submit to man, but are not tame, let alone affectionate, as many keepers know to their cost. Only animals willing to bow to human desires have even a chance of becoming domestic. As Darwin put it, 'Complete subjugation generally depends on an animal being social in its habits, and on receiving man as the chief of the herd or family.' A hundred and fifty million Indian Water Buffalo live in harmony with their owners, but that creature's close cousin the African Water Buffalo has a strict hierarchy within its herds. The African Water Buffalo is among the most dangerous of all mammals. Its pecking order does not give entry to men and the animal has never been broken in. Elephants are semi-domestic at best and kill many people each year, and plenty of breeds of cattle are happy to attack their masters. Sheep, too, show elements of the fragile personality of their ancestors, for they panic if disturbed. An aggressive animal is no help to man or beast and farms need good behaviour. Men bred from the most submissive, which led farm animals to become less alert, less active and less angry than their ancient fellows.

Their good nature is coded for within the skull. The brains of domestic animals are, without exception, smaller than those of their ancestors – the pig by a third and the horse by a sixth. The hardest job for the first farmers was to persuade the wild to become tame, but once that behavioural Rubicon was crossed the domestics could afford to put aside their need to outwit Nature and pass the job to ourselves. Many now pass their lives in a sort of calm and extended youth in which the trials of life are dealt with by their masters. The price of tolerating human company was a much diminished intellect.

Nowhere is the importance of behaviour better seen than by the fireside. The wolf is aloof, suspicious and avoids humans as much as possible. Dogs live in a mental universe quite different from that of their ancestors. Given a choice of dog or man, a wolf cub will run to the dog but a puppy to the human. Wolves hunt in packs, while feral dogs live in chaotic and quarrelsome groups that soon split up. The canine mind has been modified in many ways. Men and women follow another person's gaze. Point at an object, and all eyes will turn towards it. Dogs share that talent and if its owner indicates where a bone is hidden with a glance or with his hand, the animal runs to the right place. Wolves are baffled by the exercise. So attuned are they to their master's moods that dogs will yawn when their owner does the same.

Men become fond of their canine companions – and they return the sentiment, for when left with a stranger the dog plays less than when with its master. The bond between man and pet can last for years, and Charles Darwin himself noted how his own favourite responded to him at once after his long absence on the *Beagle*. Owners often delude themselves that their pet understands everything they say. That is not true but the animals have without doubt gained an insight into the human mind. They became fully domestic as soon as they could tolerate the other members of their household.

To gain that talent, the dog brain has been modified. Regions associated with aggression shrink and the hypothalamus, the structure that links the nervous system to the world of the hormones, is much altered. As a result, our favourite pets, unlike their wolfish ancestors, breed all year round. The hypothalamus is also a centre of activity for the nerve-transmitter serotonin, a chemical that, when in short supply in humans, is associated with aggression, impulsive behaviour, depression and anxiety. A deficiency of the stuff is also behind the attacks of rage that affect Springer Spaniels. Other fierce breeds also have low levels. High serotonin may hence be a key to the animal's calm personality. Drugs that influence its levels in humans (Prozac included) do reduce aggression in dogs as a further hint that the chemical was a key to a place around the fire.

A remarkable experiment in Russia has begun to disentangle the chemistry of calm. In the 1950s, Dmitry Belyaev became interested in the inheritance of social skills in both animals and man (a risky pastime at the time, for genetics was under attack from Stalin). He began to work on silver foxes, a coat-colour mutant of the red fox, which had first been bred in captivity on Prince Edward Island in Canada in the nineteenth century. Their elegant fur was a valuable commodity in icy Russia, but the caged foxes were almost impossible to control as they were so aggressive and fearful. After a few years he moved his experiment to Siberia, where it continues today.

At first his stock was suspicious and agitated in the presence of a keeper, albeit far less so than the animals first captured in Canada. Each generation, Belyaev chose as parents those best able to withstand the sight of a human without an attack of frenzy. The rules were strict, for just one male in thirty, and one female in ten, was allowed to pass the test and breed. Within a few generations he saw a great change. The creatures became calmer and friendlier.

They wagged their tails and learned to bark. Soon they revelled in human company, and for a time were sold as pets to raise funds. Even their appearance shifted, to embrace piebald coats, curly hair, floppy ears and blue eyes. Like dogs, but not like foxes, they have sex all year round – and like dogs they are good at the hidden food test, while the unselected foxes fail. Thirty generations on, almost all the animals are tame.

The new and tranquil beasts were cubs that never grew up. Selection for lack of fear led over the generations to an increase in the nerve-transmitter serotonin. Among the few other genes that changed were some involved in the synthesis of the red blood protein haemoglobin, which were less active in the tamed animals. That seems odd, but those proteins also help soak up certain chemicals involved in the serotonin response to stress.

In fact, the real revolution in the life of the silver fox had taken place long before the Siberian experiment began. Their ancestors in Canada, when first captured, had found it almost impossible to cope with humans. On nineteenth-century Prince Edward Island, where the 'silver' mutation was discovered, it took years before any fox could be persuaded to reproduce in captivity, let alone to tolerate a keeper. Once that sexual barrier was breached, artificial selection could begin and the rest of the emotional agenda followed. Like dogs, farmed foxes – of both the amiable and the aggressive stock – diverge from their wild ancestors in the activity of a whole series of brain genes, some of which alter transmitter levels. However, there are almost no discernible differences in the activity of a sample of brain genes between Belyaev's newly serene silver foxes and their unselected and agitated relatives in their cages. Friendliness, it appears, demands fewer mental adjustments than does the simple but formidable task of coping with human company. Most farm animals show few signs of amiability towards man, but for them, and their owners, acceptance is quite enough. To cross that barrier rescued them from the wild. The simulacrum of comradeship, as seen in tamed foxes and domestic dogs, came from later changes in other genes.

The first farmers modified their charges in both body and mind and their modern descendants do the same in a more rational fashion – but the new way of life altered the ancient farmers' minds and bodies as well. The new domestics, both plant and animal, were agents of selection upon those who had tamed them. The past ten millennia have been an era of exceptional change for human-kind, for *Homo sapiens* has evolved fast since agriculture began.

Our own genetic equivalent of the silver fox mutation, the blonde, is a creature of the fields. Just one person in fifty, world-wide, has fair hair. Before Columbus confused matters, almost all of them lived within two thousand kilometres of Copenhagen (and their close relatives the redheads abounded in Scotland, Wales and Ireland). The three or four genes responsible have become common in recent times. Farming is to blame.

North-west Europe is (or was, before the development of modern varieties) the only place on Earth where grain can be cultivated north of a line that passes, more or less, through Birmingham. Wheat, barley, rye and the rest need warmth to grow. In the Middle East, where those crops began, the sun shines upon both the fields and those who cultivate them. Our own landscape is dreary for many months of the year, and the peasants till their fields in gloom. Even so, the Gulf Stream imports energy from the tropics and heats up the ground at the end of winter, when Europe is short of sunbeams but seeds need warmth. As a result, the crops can flourish. For the five thousand years since they arrived in northern parts, a grain-based economy depended on Neptune's help.

I once spent a decade in Edinburgh and saw the sun for a few days. My present home in London has, by comparison, the equiv-alent of an extra whole month of full sunshine each year. Scotland has the worst health in western Europe and Glasgow, its cloudiest city, has levels of chronic illness higher than any other British town. Perhaps its climate is as much to blame as its much-discussed fondness for alcohol, tobacco and deep-fried Mars Bars. Vitamin D deficiency is twice as frequent in Scotland as in England and any gene that reduces skin pigment, improves the ability to soak up sunshine and make the crucial substance would be favoured. The nation has, as a result, plenty of blondes, and the incidence of the gene for red hair, with the almost translucent skin that often goes with it, rises to nearly one in three.

A cereal diet (even when its ingredients are transformed into sticky sweets) is all well and good when supplemented by other foods, but is risky when life is just one grain after another, which for our peasant ancestors it often was. Cereals are low on vitamins, vitamin D, the anti-rickets substance, most of all. A move out of Africa had, many years earlier, led to the evolution of white skin to help the northern hunter-gatherers to make the vitamin in sunlight. The arrival of farmers in the English Midlands marked a new challenge, for north of that earthly paradise the winter sun is so weak that a typical Greek, Spaniard or Italian with their dark skin and hair cannot make enough of the vitamin to stay healthy. Any child born to such a dreary diet and dank climate who inherits a new mutation for fair hair and skin is at a real advantage, for the sun can penetrate further into their flesh. The infant can make more of the crucial chemical for more months of the year and is safe from rickets. The Age of the Blonde began with the first harvest. Quite soon the homeland of the flaxen-haired expanded to overlap that of the northern cereal-growers almost precisely, with its high point in Scandinavia and north Germany, where more than half the population has fair hair, and where muesli is still a central item of diet.

Natural selection by plants acted upon the peasants in other ways. A muesli-eater digests a lot of breakfast before he swallows it, for enzymes in the spit break down starch into sugars that can be absorbed. People from places with a high-starch diet, those from northern Europe included, have up to fifteen copies of a gene for the crucial enzyme compared with just four or five in peoples who eat wild fruit, meat or fish instead. The farmers found another marvellous way to get goodness out of grain (and as an added bonus to avoid winter gloom) when they invented beer. That was bad for their brains but good for their guts, for bacteria do not like alcohol and ale was safer to drink than was the polluted water of ancient villages. Since brewing began natural selection has done its job well, for almost all Anglo-Saxons can swill the stuff down. Most Asians cannot, for they lack the bibulous West's new and potent version of the enzyme that breaks down the poison.

Animals, too, changed the fate of their keepers. Most people across the world (and most adult mammals) cannot digest milk once they have left their mother's knee because they lack the enzyme needed to do so. It works in the small intestine to break an indigestible milk sugar, lactose, into two simpler sugars, each of which can then be absorbed. In many animals – and most humans – the gene responsible is switched off not long after birth. If they drink milk as adults they feel bloated or suffer from diarrhoea. Fortunately, milk products such as butter, yoghurt or cheese do not have such an effect.

For many northern Europeans, in contrast, milk stays nutritious throughout life. Once again, natural selection has done its job: a mutation that appeared soon after cattle were tamed allows the lactose-cutting enzyme to persist and those who have it to digest milk when they grow up. Nineteen out of twenty Swedes but no more than one in ten Sicilians have that talent. The map of its distribution fits with that of genetic diversity in the local cattle breeds (which is itself a hint as to how long the animals have been on farms). Milk tolerance is most common in the homeland of the blonde, perhaps because its calcium also helps build healthy bones. Eight-thousand-year-old remains of fat from cheese or yoghurt caked on to pottery fragments from Anatolia indicate that cows were milked there at that time, although in Greece they were used only as meat. Even so, the locals probably did not drink raw milk, for DNA in the bones of Europeans from around three thousand years later still show no signs of the variant for tolerance of the stuff. Domestication led to human genetic change, rather than the other way around.

Natural selection leaves its footprints on the double helix in many ways. Long stretches of homogeneous DNA on either side of the European genes for blonde hair and milk digestion show that the new and beneficial variants dragged their neighbours along as they swept through the population over the past few thousand years. Such stretches of homogeneity are a hint of the action of selection – even if in most cases we have no idea of what gene was involved or why. In time, the segments are broken up by the reshuffling that accompanies sex. A search through the DNA of people from Africa, Asia and Europe reveals many such segments, each a relic of a sudden attack of selection – often since the origin of agriculture. The change in lifestyle and diet in the ten millennia that followed caused much more evolution than in an equivalent period during the millions of years that humans lived as hunters since the split from chimpanzees. Man, like his animals, has changed a lot since he moved to the farm.

For mankind, domesticity itself began long before he began to till the soil. The notion of *Homo sapiens* as a house-trained ape has a long history. Darwin himself saw the parallels between the farmyard and the parlour: 'We might, therefore, expect that civilized men, who in one sense are highly domesticated, would be more prolific than wild men . . . The *increased fertility* of civilized nations would become, as with our domestic animals, an inherited character.' He had hoped to add a whole chapter on humans to his work on the origin of farm animals, but he saw that the book was already 'horridly, disgustingly, big' and abandoned the idea. That chapter has now been written.

As in the famous Siberian foxes, the real revolution in the human line took place when an ape became human. There have been further and more minor adjustments as hunters settled into their new life in the fields. Many of the physical changes in the human line since it emerged resemble those found in domestic animals. Compared with our ancestors, we have a lighter build, thinner skull, shorter jaw and smaller teeth, and with less marked differences between males and females. We quarrel less about sex and are less enthused by it than are our closest relatives and, like dogs, men and women copulate all year round, rather than in a short season as do wolves. Our breasts – like the cow's udders – are larger and milkier than theirs. Like pigs, we store fat more readily than do our great ape kin and are less keen on physical activity. As is the case for dogs, sheep and cattle, various odd physical mutations (blonde hair, light skin and blue eyes included) have emerged in some populations, although we have not yet gained a patchy coat. Our brains, alone, have not diminished.

Such changes in physical structure, together with those towards baldness, an upright gait and an agile mind which marked the transition from ape to man, involved a large cost, the death or sexual failure of millions who could not cope with the new way of life. The more recent changes in skin colour, the ability to drink milk or beer, or to digest grains demanded just the same sacrifice. The speed at which advantageous genes have spread since the origin of farming suggests that the price was high indeed. The same process is at work today. We face a new abundance, quite different from anything in the evolutionary past, and have not yet evolved to deal with it. We may do so, but the process will not be cheap. For most of history, humans have had to cope with shortage rather than excess and have evolved mechanisms to guard against excessive weight loss when food is short. Our bodies deal less well with today's glut. Starvation disguised as surfeit means that evolution's inexorable machine has cranked up again, with natural selection by diet as active as it was ten thousand years ago.

Our individual response to excess depends strongly on our DNA. A study of twin boys and girls suggests that around seven-tenths of the variation in body weight within a population is due to genetic variation. Identical twin children also resemble each other in how much they will eat if offered a huge meal. Adult twins paid to gorge themselves, or to starve, for several weeks also tend to gain, or lose, weight – to resist, or to surrender to, the challenges of the new diet to the same extent.

Dozens, perhaps hundreds, of genes have been blamed for the new wave of obesity. Fatness runs in families but so do frying pans – and fat cat-owners tend to have fat cats. Their pets share their diet but not their DNA. Nature and nurture work together and the inheritance of pot bellies is – like that of almost every-thing else – not simple. The notion that fat people can blame the way they are born and ignore what they choose to eat is wrong. Instead, like alcoholics they are more at risk of a certain kind of diet than are others and must struggle harder against temptation. Many people fat today would have been thin a century ago, what-ever the nature of their DNA.

Even so, genes have a real influence on waistline. Most of those that predispose to obesity have a small effect, but a certain variant, when inherited in double dose (as it is by ten million Britons), increases body mass by three kilograms above average. Even a single copy, as borne by almost half the inhabitants of these islands, adds a kilogram. The DNA variant concerned changes appetite and is active in parts of the brain involved in hunger or satiety. The gene is harder at work in starved individuals than those who have just eaten. It makes no difference to their weight at birth but babies with two copies begin to pile on the kilo-grams within just two weeks. Other genes that dispose to obesity alter the efficiency with which food is soaked up or the rate at which the body burns its fuel.

The environment itself has effects that stretch over the genera-tions. Just as alcoholic mothers have damaged babies, women who eat fast food have fat children not only because they pass on their genes, but because they were themselves overweight while pregnant. About a third of all pregnant Americans – and half of pregnant African-Americans – are obese. Their internal economy shifts to deal with the problem, as does that of their unborn child. The evidence is clear, for those who lose weight, perhaps through stomach surgery, between one child and the next, tend to have lighter babies than before. The foetuses of overweight mothers respond to the high levels of insulin (the hormone that controls blood sugar) in their mother's bloodstream and are born attuned to lay down fat. Genes that dispose to diseases of the obese are then put under further pressure.

The biggest threat to the overweight is diabetes; not the rare variety that affects a few infants and can be treated with insulin but a related illness that comes on later, defies treatment and has a strong tie with an expanded waist. The problem arises from a resistance to insulin. Its symptoms include heart disease, kidney failure, blindness, nerve damage and even gangrene. It was once a disease of the elderly, but is seen more and more in children and adolescents. Just one extra notch on the belt adds a lot to the danger and those in the top tenth of trouser size have twelve times the risk of diabetes than do the slim.

Half a billion people will soon suffer from adult-onset diabetes. Unless matters improve, a baby born in the USA today has a one in three chance of the condition when it grows up and the illness already takes up a sixth of the country's entire health budget. Even in Britain, two million people show signs of it. In some places the figures are dreadful, with half the adult population of the island of Nauru, in the Pacific, afflicted. Genes that increase body weight are most to blame, although others do increase the risk in both the fat and the thin.

Obesity is in part inherited, and is a target of natural selection because it kills many people before their time. Its effects on the evolutionary future are made worse because those who suffer from it face not just premature death but sexual failure. Fat people tend to have fewer children than average. Apart from the romantic problems involved, obese men find it harder to sustain an erection, and obese couples copulate less often, than do the fashionably slim. Even worse, a fat man's sperm count drops by around a quarter, perhaps because his over-insulated testicles are too warm. Female fertility, too, drops with every extra kilo. Excess fat interferes with the menstrual cycle and has other harmful effects. Among women anxious to become pregnant even a slight weight excess increases the time before a favourable outcome by a month, and it takes nine months longer for an obese woman to have a good chance of becoming pregnant than for a person of normal size. In addition, overweight women are more liable to miscarry and their children are at higher risk of birth defects.

All this means that natural selection by diet is once more hard at work, as it was when agriculture began. Darwinians, faced with the problems that have emerged from the new way of life, can hence afford a certain grim optimism about the future. Man evolved to deal with a changed diet in the first food revolution, and will no doubt do so in the second, whatever the cost. In this era of global glut, natural selection may act on future generations until they return to slimness and health in an affluent world, just as the descendants of the first farmers evolved their way out of their dietary problems.

The crude tools of evolution are, needless to say, far less effective in moulding the future than is the simple human ability to learn from our mistakes. Societies facing the waistline problem are better advised to consider the risks, plan ahead and eat less than to await the attentions of biology. Everywhere, people are exhorted to improve their diet and take up exercise although so far the propaganda has not been particularly effective. Even Marie Antoinette was trying to help. The famous 'cakes' offered to her starving nation were not rich and lard-laden delicacies, but baked crusts that might otherwise have been thrown away. A simple error gave rise to a legend of political incompetence and to a sticky end. In these days of excess, her regal counsel seems more sensible than it did at the time of the French Revolution. Whether people will take her advice and modify their lethal habits or whether they will wait for natural selection to do the job, it is, to quote Zhou En-lai on that interesting political event, too early to say.

CHAPTER VI

THE THINKING PLANT

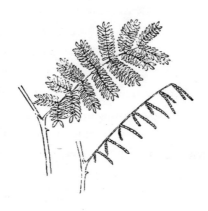

Deep in the Amazon jungle a creature snakes into the light. As it climbs cautiously through the branches it senses a brighter spot on a distant tree. After weighing up the risks of abandoning its present post it plunges back into the gloom of the forest floor and creeps across the ground until at last it reaches its target, scrambles upwards and triumphs to bask high in the tropical sunshine. The vine – for such it is – shows every sign of foresight in its behaviour. The notion that a plant might act in what appears to be an intelligent way is alien; less so than before time-lapse films speeded up the circling of shoots or the opening of flowers, but unexpected at least. Can such a simple creature really plan ahead?

Romantics have long been convinced that the vegetable kingdom has a mind of its own. Gardeners talk to their crops in the hope that they will flourish, while tree-huggers, when not in close embrace with a trunk, often play a part in the conservation movement. Real enthusiasts for botanical intelligence believe that cacti grow fewer spines when they listen to soft music and put them out again when they see a cat. The Japanese even enter into two-way conversations with their green friends. They have patented an electronic device through which a flower can chat to its owner or, when thirsty, ask for water. In the 1920s, the famous Indian physicist Chandra Bose, a pioneer in the study of electro-magnetic waves, worked on electrical activity in plants. His subjects did generate a measurable current when damaged (an observation that led to genuine scientific advances) – but Bose was also certain that music and kind words could set off the response.

Dubious as such claims might be, the mental universe of plants is, if nothing else, useful fuel for metaphor. Shelley writes of a garden in which a mimosa droops in response to a rejected lover's despair: 'Whether the sensitive Plant, or that/ Which within its boughs like a Spirit sat,/ Ere its outward form had known decay,/ Now felt this change, I cannot say.' The Latin name for Shelley's sympathetic subject is *Mimosa pudica*, in reference to its bashful nature, and the Chinese call it 'shyness grass'. Whatever the plant's mental state, it does respond to the outside world. For most of the time, a mimosa's branched leaf stands proud, but a slight touch, or a gust of wind, causes it to droop in a hang-dog fashion. It can take hours to recover. At night, no doubt exhausted by the emo-tional turmoil of the day, the leaves close up and their owner goes to sleep.

Shelley's lines are both a literary device and an accurate obser-vation. They also say something about the relationship of mind with brain. If a mimosa can act in what seems a rational way even in the absence of any hint of cerebral matter, what does the end-less debate on that topic mean? Philosophers, like poets, should perhaps pay more attention to botany.

Charles Darwin, as a competent scientist, had no real interest in such metaphysical ideas (he did, admittedly, claim that plants sometimes recoil in 'disgust'). He was nevertheless curious about their ability to react to the conditions in which they are placed. He wrote two books on the subject. *The Movements and Habits of Climbing Plants* of 1875 deals with how ivy, brambles and the like find and scramble up their vertical helpers. *The Power of Movement in Plants,* published five years later, asked wider and more radical questions about how all plants respond to the outside world. It had, he wrote, 'always pleased him to exalt members of the botanical world in the scale of organised beings', and in those volumes he succeeded. Together, the two books discuss three hundred species. Darwin placed the plant kingdom on a higher scientific plane than ever before, for the experiments in his greenhouse laid the foundations of modern experimental botany.

His home county was in those days famous for hops. So fond was the British working man of his beer that Kentish fields were filled with poles and wires up which the bitter vines were trained. Each September tens of thousands of labourers and their families came from London to pick the crop and to have what, in Victoria's glorious days, passed as a holiday. *Climbing Plants* asks a simple question. How does a hop find a support and climb up it?

To succeed, its shoots as they peep above the soil must seek out an upright of the right size even if they emerge some distance away. Then they must twine around it to clamber upwards. The talents of the hop were the introduction to a new world of botanical behaviour.

Most of the work was done with the help of Darwin's son Francis. It was, as ever, interrupted by ill health: 'The only approach to work which I can do is to look at tendrils & climbers, this does not distress my weakened Brain.' Charles noted, first, that a pot plant in his sick-room circled round as it grew. He and Francis began to cultivate a variety of species beneath clear glass plates upon which the position of the tip could be marked with ink. They saw that the shoot of a young hop travels round all points of the compass. On a hot day a complete revolution took about two hours. Should the questing tip touch a pole, the hopeful climber then changed its behaviour, snaked around it and found its way to the top. What looked like forethought depended on just three simple talents: the ability to circle, a sense of touch and the capacity to tell up from down.

Father and son went on to study other plants that climb not just with their shoots, but with structures such as tendrils, hooks or adhesive roots. Whatever the details, almost all the climbers gyrated until they found a support and, once found, clambered away from the ground. The Darwins soon discovered that all shoots, even in species that do not climb, in fact circle to a greater or lesser degree. In the same way, all plants can modify their growth to avoid an obstacle, and all can sense gravity. A hop's unusual powers depend – as do many patterns of animal behaviour – on natural selection's ability to modify talents that already exist.

The second book, *Movement in Plants*, went further. It describes experiments on the sensitivity of roots, shoots and more to light, gravity, heat, moisture, chemicals, touch and damage. The research was far ahead of its time. Although they did not invent the name, father and son discovered the first hormone – not in animals (an event which had to wait for almost thirty years before scientists at University College London found a chemical messenger in the blood of dogs) but in plants. So impressed was Charles Darwin by the powers of shoots and radicles (the first structures to emerge from the seed at the time of germination) that his book ends with a dramatic claim: 'It is hardly an exaggeration to say that the tip of the radicle thus endowed, and having the power of directing the movements of the adjoining parts, acts like the brain of one of the lower animals; the brain being seated within the anterior end of the body, receiving impressions from the sense-organs, and directing the several movements.'

Any creature, animal or vegetable, needs, as it copes with the outside world, to find out what is going on, to pass the informa-

tion to the appropriate place and to respond to the challenges presented by Nature. Men and women do the job with eyes, ears,
nerves, brains and muscles. Plants have none of those, but cope
remarkably well – and in some ways they put our own abilities
to shame.

Why climb? Lord Chesterfield got it right. In one of his notorious letters of advice to his son he wrote that 'A young man, be
his merit what it will, can never raise himself; but must, like the
ivy round the oak, twine himself round some man of great power
and interest.' A plant with such an ambition uses its support to
reach a lofty place to which it could otherwise never aspire. The
helper might come to regret its generosity, but the advantages
from the social climber's point of view are clear.

Such behaviour opens up a new universe of opportunity. The
plants that first evolved flowers able to attract pollinators, and
those that first developed fruits to persuade animals to move seeds,
each discovered a whole new set of habitats and a variety of ways
of life. As a result their descendants burst into a variety of form.
The ability to climb is less dramatic than are fruits and flowers, but
those who take it up have also evolved into a vast diversity of
kinds. A hundred and thirty different families in the botanical
world have climbers. Within each group, those agile creatures are
represented by many more species than are their earthbound kin.

Birds, bats, flying squirrels, snakes and fish all take to the air but
in different ways, with modified arms, hands, bodies or fins. In
the same way, plants have called upon different organs to help
them climb. Some, like hops or peas, use tendrils, based on stems
or leaves. Others, such as clematis, have altered leaves in other
ways, or evolved specialised roots or hooks that allow them to
scramble. Roses have hooks. The ivy uses roots to clamber fifty
metres and more up cliffs, houses or trees while Virginia creepers
go to the opposite extreme and use shoots. In a certain group of
ferns the fronds grow around the support to make their way
towards the light.

The habit is ancient indeed. Three hundred million years ago, the Earth had vast coal swamps filled with fern-like trees fifteen metres high. The forest had plenty of vines and climbers, which used structures like those of modern plants to struggle into the sun. It became a tangled and impenetrable mass until at last the whole lot was wiped out as the climate changed.

Tropical forests are still the capital of the scramblers. There, every plant must fight to reach the sunshine against thousands of others. Many jungles are filled with lianas, woody vines that loop down from the trees. In most places they represent less than a tenth of the total mass of live material – but their tactics are so effective that their leaves fill half the canopy. Almost half of all woody species in the Amazon basin are climbers, with fifty or more different kinds in every hectare. They are fond of gaps, places left open when a moribund giant crashes to the ground or when farmers clear a space (which is bad news for the farmers themselves as they compete with the creepers to grow a crop). When forests – tropical or temperate – are broken up by loggers, the lianas and their relatives thrive even as the trees upon which they depend are destroyed.

Climbers climb, in the main, to get into the light. Another good reason to take up the habit is to escape, like a baboon pursued by a lion, from ground-based enemies. Leaves near the surface get chewed more by slugs, snails and the like than do those up in the air. In the arid deserts of northern Chile, convolvuli often grow near cacti or thorny shrubs. After an attack by hungry mice, or by scientists with scissors, they at once increase the rate at which they twine and put out more tendrils in the hope that they will reach a shrub and clamber into the safety of its spiny branches.

Darwin noticed that most twiners needed a rather slender pole if they were to make progress – British climbers, indeed, never curl around trees. The upright must also be rough enough to give them a chance to hold on. The climber does not cling with its whole length, but sets up a series of contact points as it moves onwards. Rather like a bloodhound, it sniffs the air now and again as it tracks its route. As it moves, the tip is raised, circles round and comes back to the stem a few centimetres further on. The details vary, with some tendrils set like a coiled spring to twist within seconds around a support as soon as they touch it. Engineers have worked out that for a smooth pillar the climber cannot manage to ascend a support more than about three times its own thickness – a twig, or a vertical wire. The rough bark of a tree makes the job rather easier. Part of the spiral motion as a hop moves on comes from an increase in the rate of growth on one side of the plant compared with the other. In addition, cell walls on that side become looser, bulge up and force the shoot to wind round and round. In time, a tendril can coil in upon itself and grow hard and woody, to lock its support in a fierce embrace.

In some species the young stems are rigid and grow upright without help for several metres – but once they touch a tree, they pounce. No longer do they need to invest in solid – and expensive – wood. Instead they become thin and flexible and begin to clamber. Certain lianas grow a flexible stem to find the open air, but once they reach a sunny spot they generate huge trunks that swing downwards from the heights and find another plant to use as a support. That noxious North American the poison oak grows as a solid two-metre shrub when it stands alone, but extends ten times higher if it can find an upright. Many other kinds take advantage of a helper when they get a chance, but stand on their own feet (or roots) if they do not. In a tropical forest, young trees of species not often thought of as climbers grow slim and tall as they lean against their neighbours. If that choice is not available, they stand alone and take up an independent life.

In many climbers, some branches have small leaves and move in a wide circle in search of a new gap through which a shoot can insinuate itself. Those that sneak through and find the sun then grow larger leaves that soak up energy. As the stems spiral away from the ground, they develop wide vessels through which to suck up water and food. The liquid has to travel through a passage many metres long, which makes drinking expensive and forces the plant to reduce water loss with waxy leaves and impermeable stems.

A tree pays a high price for its parasites, for they suck water and minerals from the soil and shade their host from the sun. West African trees in the presence of lianas grow at no more than a fifth of the rate of their fellows. A few climbers can kill. The strangler fig, once it has reached the canopy, sends roots down from its eyrie. As they grow, the aerial roots wrap themselves around the supporter's trunk, fuse together and squeeze it to death. The lethal tenant is left vertical and proud with its own roots in unencumbered soil. In other trees the benefactor crashes to the ground under the weight of its visitor, but by then the fellow-traveller has moved on in the canopy to bask in sunlight at a second tree's expense.

Some plants twine clockwise and some anti-clockwise – as in the famous case of the right-handed honeysuckle and left-handed bindweed. A mutation called 'lefty' in a small mustard plant persuades the normally straight stem to spiral to the left, while another causes a bias in the opposite direction. Each changes the shape of a crucial protein in the cell skeleton. The molecule looks like a string of asymmetric dumb-bells, with each element lying together head to tail to form a helical and hollow cylinder. The mutations enlarge one or other end of the protein and deform the cylinder, which changes the pattern of cell division and causes its owner to twist. In an echo of the Flanders and Swann song, plants with a single copy of each mutation do indeed grow straight up (although they do not fall flat on their face). For reasons unknown, a bean that normally circles to the right increases its yield if forced to twist in the opposite direction.

Climbing plants are of interest to gardeners, to brewers and to wine-drinkers but for Darwin they were an introduction to a whole new range of botanical talents. *Movement in Plants*, his second volume on the topic, shows that leaves, root-tips, shoots and other parts of all species, climbers or not, are in constant motion. They respond to circumstances in much the same way as do animals. Plants might be slower, but they get there in the end.

The hop's ability to climb is matched by the skills of every seedling as it emerges into a hostile world to fight for light, for water or for food. *Movement* contains a graphic description of what might appear to be the purposive actions made by a new-born plant in its first days. In the struggle to turn into the right position, to push its root into the soil and its shoot into the air, a seed as it germinated reminded Darwin of a man thrown on his hands and knees by a load of hay falling on him. 'He would first endeavour to get his arched back upright, wriggling at the same time in all directions to free himself a little from the surround pressure . . . still wriggling, would then raise his arched back as high as he could. As soon as the man felt himself at all free, he would raise the upper part of his body, whilst still on his knees and still wriggling.'

To escape to safety the shoot and the root must each respond to light, to gravity, to touch or to other stimuli. We ourselves live in a universe of senses – of sight, sound, smell, taste, touch and, the for-gotten sense, position. Seedlings have no noses, tongues, fingers or ears, but they too perceive the outside world. Animals use electric-ity and chemicals to pass messages through the body – and so do plants. They have no muscles – but they grow to where they need to be, or move with the help of molecular machinery quite like that which drives our own limbs. As Darwin put it, 'it is impossible not to be struck with the resemblance between the . . . movements of plants and many of the actions performed unconsciously by the lower animals . . . the most striking resemblance is the localisation of their sensitiveness, and the transmission of an influence from the excited part to another which consequently moves'.

Without eyes, ears or nerves, how can a plant know which way is up, what has touched it or whether the sky is blue or grey? Now, we have begun to find out.

Father and son identified two general kinds of activity – those in which just a response, and not its direction, is related to the external trigger and those that involve a move towards, or away from, an outside stimulus. Among the former, they noted that many plants open and close their flowers in sunlight and shade, or 'go to sleep' as they fold their leaves at night, perhaps to reduce the amount of heat lost by radiation. Some, like the mimosa, also responded to a sudden prod with a collapse of the leaves in an attempt to frighten off a hungry insect, or to expose an enemy to the thorns with which its branches are decorated. All those with sensitive leaves slept at night but plenty of the sleepers were quite indifferent to a poke.

For the mimosa and its fellows such actions come from a sudden loss of internal pressure in each frond, which spreads to the leaves next to that actually touched. Certain cells held in a bulge at the base of the leaf-stalk are crucial, for if they are rubbed or tickled they act as hinges and the leaf folds at once. They are more sensitive than are our own fingers. The hinges also control the response to darkness to light. Each has a long hair that acts as a lever and is embedded in a sensory cell. On a stimulus, the magnified movement at the base of each sets off a response in the hinge as charged molecules are pumped across its membrane. At once, water is lost, parts of the internal skeleton of the cell collapse and the leaf folds up. In time, the plant forces water back into the crucial structure and sets it ready to respond to the next challenge. The pattern of two opposed forces at work to close or to open the leaf is rather like our own arrangements, in which one muscle causes a limb to extend and another makes it flex.

Many flowers can tell the time and the ancients sometimes set the hour of prayer with a quick glance at the garden. The great classifier of life Linnaeus even designed a bed filled with blossoms that opened at different hours to make a crude botanical timepiece. The talents of many such species – such as the sunflowers that track the sun through the day – turn on no more than a direct response to outside stimuli. Mimosas have a more subtle sense of the hours, for when placed in constant darkness the rhythm of sleep and wakefulness persists. They have an internal biological clock, independent of light and dark. The plants undergo what Darwin referred to as 'innate or constitutional changes, independent of any external agency'.

An internal timer, based on the build-up and breakdown of chemicals, maintains the daily rhythm. The clock is not precise and will wander away from strict time if kept in constant light or dark. Different species have internal timers with a cycle that varies from around twenty hours to about thirty. Dawn resets the mechanism, which hence keeps up with the shifts in hours of daylight as the seasons progress. The inner and the outer world interact, for in mimosas the leaves do not fold up at night unless they have been illuminated during the day.

Such movements have what might look like purpose, but they lack direction. Other botanical talents give the impression, almost, of having a definite goal in mind. A plant's life is ruled by the sun, by water, by food and by predators. To survive, it must avoid its enemies and find its friends and, like an animal, hunt for food, water, shelter and – most of all – sunlight.

The Darwins discovered that young shoots will grow towards even a dim light. That simple observation led them to their most significant result: the discovery of an internal chemical message – a hormone – that altered growth. It was the first of thousands of such chemicals now known.

Their experiment was simple but ingenious. A shoot of grass bent over towards the light. It did so, they found, only if the beam hit its topmost point from one side. If the very tip was covered, or the light was directed to a spot just below it, the shoot remained unmoved. In addition, when the plant was buried in sand with only the tip left in daylight and the rest in pitch blackness, the buried shoot grew towards the source of illumination although the rays never touched it. Short bursts of light had about the same effect as did a single longer glow and even very low levels did the job. The tip of the shoot, they realised, was sensitive to the sun's rays and somehow the information on where it came from was passed ('influence is transported') to the stem below and per-suaded it to alter its activity.

Years later, in 1913, came direct proof of a chemical messenger. The amputated tip of a stem was placed in daylight on a piece of soft sponge. The sponge soaked up the crucial substance as the scrap of tissue pumped it out and, when it was laid on to a cut stem whose own tip had been removed, the shoot at once altered its pattern of growth. The botanical envoy was named 'auxin' (after the Greek *auxein*, 'to grow'). It was the first hormone.

For plants and animals alike, to learn about the world outside is not enough. To respond to the messages of opportunity or danger that pour in, information must be transmitted from the point of arrival to a body part that can respond to them. News about the outside world travels through an animal's body through many routes. Nerves pass it on from cell to cell (and all cells, nerves or not, talk to their neighbours), while distant tissues communicate with the help of chemicals released into the bloodstream. Plants have no nerves, but they, too, pass instructions between cells through spe-cial channels that traverse their thick walls and allow the living parts to touch. Darwin's hormone travels downwards from the shoot tip in that fashion and the channels in addition transfer proteins, nucleic acids, hormones and even viruses. Plants also have open vessels – but unlike our own bloodstream, liquid does not circulate but moves from roots towards shoots or leaves, whence it is lost by evaporation. As a result, any flow of information is one-way. The hormones that travel through the vessels include proteins and molecules that control cell division and cell death as well as others that control the decision to flower or to store food. Other signals tell the dark world beneath the soil when spring has come while yet more keep an eye on the balance between food and growth or send warnings about the arrival of an enemy.

Dozens of plant hormones are known. The chemicals resound through their tissues in response to light, heat, damage, the passage of time and more. Some emerge in unexpected places. Human urine applied to a decapitated shoot alters its growth because a plant's auxin passes unchanged through the body of those who eat it (and the substance was in fact itself first purified from that invaluable fluid). Now the messengers are studied not just with chemistry but with mutant plants whose altered growth is due to an aberrant response to hormones.

Most plant hormones are simpler and smaller than our own. Some have a chemical structure based on closed carbon rings, but many are small proteins. A few even look rather like the steroids that control human sexual attributes. Like mammalian chemical messengers, they are often arranged in pairs, with some that promote an action and others that oppose it. Each has a receptor on the target tissue to which it binds, and each acts – as do those of animals – to stimulate or repress the action of particular genes. Some cause individual cells to expand or to contract, while others change the rate of cell division – should, for example, cells divide faster on the dark rather than the light side of a shoot, then the whole structure bends towards the source of illumination.

Such molecules determine when their masters will ripen, lose their leaves, move towards or away from light and gravity, fight infection, and more. Those concentrated in the tip of a shoot act to suppress the activity of sections of the plant that lie below them. Auxin diffuses downwards and prevents the growth of buds that might compete with the tip itself for light. Cut off the tip and those segments burst into life – which is why gardeners prune their fruit trees to get a dense bush.

The locks into which the auxin keys will fit have also been discovered. Many of the inherited changes in shape treasured by
gardeners result from errors in the hormone genes or their receptors. The auxins persuade the shoot to grow up or the root down,
the flower to bloom and the fruit to swell under the influence of
auxin from its seeds. The sinister movements of the sundew as it
rolls its leaves over a trapped insect are due to another member of
the same chemical family. In some ways the auxins resemble the
substances involved in nerve transmission more than they do
animal hormones. Indeed, some of our own nerve transmitters are
found in plants but quite what they do is not yet clear.

The auxins and their relatives have often been turned to useful
ends. Gardeners and farmers use artificial versions to help cuttings
to take root. Agent Orange – so named after the colour code on
its barrels – was an artificial auxin that caused vegetation to grow
itself to death. It was used for a decade at the time of the Vietnam
War. Over seventy-five million litres were used in an attempt to
destroy the guerrillas' crops and to open up the forest to expose
the enemy. Its military value was never proved and the chemical
was abandoned when found to be contaminated by the poison
dioxin. Even so, synthetic auxins are still used as herbicides and
appear to be safe.

The leaves that hid the Viet-Cong from aircraft formed a dense
screen as the trees that bore them struggled for life – and for light.
All plants need sunlight and will fight for that precious resource,
often to the death. Every forest is the result of a silent battle
between the leaves far above, as each jostles to get a view of the
sun. Together, they block its rays. Some bathe in its beams, but
others fail, pale and die. To survive they need to pick up the solar
radiation, measure its intensity and move, or grow, in response.

The Darwins found that shoots can pick up light and pass on
information but they had no idea of how they noticed the solar
presence or of their exquisite sensitivity to wavelength, intensity and

direction. They do the job in several ways, with a variety of special molecules, some of which have equivalents in the animal kingdom.

One group of receptors known as phototropins picks up blue light and plays a large part in the growth of shoots towards a source of illumination. Another group, the phytochromes, is sensitive to the longer waves, the red and infra-red. Phytochromes have a protein skeleton matched with a second chemical structure based on linked rings of carbons bent into a molecular knot. The molecule is poised like a set mousetrap and when light strikes it it flips from one shape to another. In the dark or shade the change is reversed. The balance between the alternative forms tells the plant how much light is in the sky. The light-sensitive part looks rather like the chlorophyll found in leaves, and – less predictably – resembles the breakdown products of our own red blood pigment (which is why jaundiced babies can be helped with a dose of intense light).

In a dense forest the proportion of infra-red that hits the ground is no more than a twentieth of that experienced by a plant exposed to full sun, because most of the energetic radiation is soaked up by the leaves above. Because the pigment measures not just light intensity (which changes as the day wears on), but also the ratio of red to infra-red, it can distinguish a shortage of light in the evening or on a cloudy day (when the proportion of the two wavelengths does not change) from an attack of gloom that arises because other leaves have shaded out the solar disc and have stolen the most valuable part of its spectrum. Once the infra-red alarm has sounded, those in the shade must respond to the emergency before the source of energy is blocked altogether.

A plant in this predicament shifts its whole pattern of life. It grows faster and the stalk stretches higher. Each leaf moves to present a flatter surface to what light is available. If the shortage continues, the leaves become thinner and more transparent and their parent becomes less branched as it reaches for the sky. If all this fails, and the light still stays red, the unpalatable truth becomes clear. The victim flowers as soon as it can to give at least some hope that its genes will be passed on before it dies in darkness.

The phytochromes are smart, but other pigments are even smarter. A third set of sensors, the cryptochromes, respond not to red, but to blue light. In a further parallel to our own eyes, with their three distinct receptors, a green-sensitive pigment has also been found. Cryptochromes have a structure rather like that of the enzymes that cut and splice DNA. They measure the intensity, rather than the wavelength, of light. Their main job is to sense how long each day is — an important factor when deciding whether to make flowers or fruit as the seasons move on. They are also hard at work in the newborn seedling as it wriggles its way towards adulthood, for they shift the whole biochemical economy of a young shoot from a life based on the gloomy world of the soil to a career bathed in sunshine.

Many of the sensor molecules have relatives in our own eyes. They too help to work out the length of each day and to assess wavelength (or colour, as we call it). Some of the magic proteins are the very stuff as dreams are made on for not only do they cause the sensitive plant to droop but they control human sleep rhythms. A long trip in a jet plane leads to unpleasant side-effects — and the cryptochromes help to put them right, which is why a global traveller finds it harder to adjust to local time in gloomy London than in sunny Sydney. Mice in which the blue-light gene has been damaged by mutation sleep more and have less active brains as they snooze — and, for reasons unknown, also show shifts in their response to anti-cancer drugs. The levels of such chemicals in our own eyes are also tied to the annual swings of mood familiar to those with seasonal affective disorder, the Black Dog of winter (although no fit has yet been found between genetic variation in the human genes and liability to that unpleasant illness). The sensitive plants droop in gloomy weather with the help of cryptochromes and so, it seems, do we.

Roots, in contrast, prefer darkness and make a real effort to avoid daylight. Once again, blue light does the job, with its own special receptor molecule in the tip. The root has another talent which helps it delve into the soil, for roots can sense the force of gravity. For climbers, too, the Earth's attraction is important, although they prefer to move in the opposite direction. Darwin found that the crucial sense organ for gravity resided in the tip of the root and the shoot and that to cut off that tip much confused the growing plant. Now it has been tracked down – and, once again, it has some uncanny similarities with the human system that does the same job.

Men and women maintain their equilibrium with a set of liquid-filled tubes in the inner ear, arranged in three dimensions, left and right, forward and back or up and down. They contain a liquid that washes back and forth as we stand, sit or move about. Tiny grains of calcium carbonate rest on special cells on the inner surface of each tube and shift as gravity or acceleration directs them. The movements of the fine hairs on each cell are translated into electrical messages to the brain to give us a sense of where we stand.

Plants do the same with special cells in roots and shoots. Each contains small grains of starch which, like the minute particles within our ears, shift as their owner moves. Mutants unable to make starch lose both their sense of gravity and the ability to circle. Darwin speculated that the questing movements he found in hops and the like depend on the Earth's attraction, but he was not altogether right, for, in an experiment that would have flabbergasted him, the tips of plants held in weightless conditions on the Space Station continue to make their measured rounds.

A closer look at both people and plants shows further parallels in the gravity sensor. In the ear, a molecular rack and pinion uses a pair of proteins that play a part in muscle to pick up the movement of the small grains as they are washed back and forth. In the plant a pair of almost identical molecules does the same job.

Poets, mystics and romantics often imagine the vibrations of a sixth, seventh or eighth sense (although Shelley had more sense than to do so). One popular candidate is magnetism – a topic tarnished from its earliest days when the German mountebank Franz Anton Mesmer claimed that 'animal magnetism' – the supposed ability of some people to open blocked bodily channels in the afflicted – could cure blindness and more. The idea was used by Mozart in *Così fan tutte* but blown out of the water by a French governmental commission headed by Benjamin Franklin. There are still plenty of magnetic therapists, who sell hundreds of millions of dollars' worth of magic bracelets, insoles and blankets in the United States each year.

Biology has gained a renewed interest in our interactions with the Earth's magnetic field. The subject still attracts odd claims: some say that blindfold students can find their way home thanks to a supposed internal compass, while aerial shots hint that cattle tend to line themselves up to face north or south. A strong magnetic field does spark off brain activity, but what relevance that has to daily life is not clear.

Many creatures do have a strong and unexpected ability to sense direction, with the help of a magnetic compass. Migratory birds, for example, have iron-rich cells in their brains, and use the Earth's field to find their way back and forth across the globe as the seasons change. To do so, they use the products of a gene remarkably similar to that of a plant's blue-light sensor. Its molecule is an essential part of an internal timer employed by the migrants as they measure the angle made by the sun as it sweeps across the sky and use the information to orient themselves north or south. In addition, it helps the birds to navigate by the Earth's lines of magnetic force.

They can pick up the field only in daylight, and the blue sensor is what allows them to do so. It works at the atomic level. Electrons come in pairs that spin around the nucleus in opposite directions. As a result they cancel out each other's ability to act as magnets. When energy – from light, heat or chemical reactions –

enters the system, some particles are knocked off balance and are left with just a single spinning electron. Such 'free radicals' need to marry another partner as soon as they can. To find a match they must change their direction of rotation. A magnetic field at right angles to the spin makes that task harder. As a result, a bird can use unpaired electrons as a compass needle that allows them to sense the Earth's magnetism. The cryptochromes generate lots of free radicals when they pick up the energy of blue light and transmit it into the cell and with their help the happy migrant gains an improved sense of direction.

Magnetism also has an effect on plant growth. The starch grains in the tip of a shoot can be moved with a magnet and the field then helps tell a plant where it stands (an attempt to test the results of that experiment without gravity perished in the *Challenger* disaster). A magnetic field also slows the growth of shoots – but only in blue light. Mutants that lack the relevant sensor are quite indifferent to its presence. Perhaps all this hints at a forgotten shared sixth sense, in both animals and plants. Darwin often spoke of 'fool's experiments', ideas that were most unlikely to work but were worth a try. He speculated on the role of magnetism in animal navigation but he would be amazed to find that it applied to the movements in the other kingdom of life. The experiment was too foolish even for him.

In another instance of what his son called his 'wish to test the most improbable ideas', he persuaded Francis to play the bassoon to a mimosa to test whether it would respond. Sweet music, like kind words, had no effect; and a later experiment in which pop music was blasted at them for hours also left them unmoved. Even so, plants do hold conversations. They use not sound to do the job, but scent. A series of chemical messengers have evolved to help, and together they provide a sense of smell to add to those of sight, gravity and the rest.

Many pump out a simple gas called ethylene, which causes them to grow faster and also helps fruit to ripen. In dense vegetation, the gas persuades leaves to struggle harder towards the light, for it reveals the presence of competitors. The ability to smell can be used for more sinister ends. Dodder, also known as devil's guts, witch's shoelaces, hellbine and the like, is a parasite related to the morning glory. It is a pest of carrots, potatoes, clover and garden flowers. Its leaves are tiny and contain almost no chlorophyll. The seeds can survive for ten years. Once they germinate, the shoot pokes above the soil and searches for a potential victim. It circles round until it succeeds. Then it inserts fine needles whose cells fuse to those of its quarry. They suck out vital fluids and the plant loses its own root, to live as a parasite. It has a strong preference for juicy species like tomatoes over tougher kinds such as wheat.

The dodder, like a bloodhound, sniffs out its prey. When allowed to germinate in a closed container with two tunnels, one that leads to a chamber with healthy tomatoes and the other to a similar space with wheat, it directs its growth towards its preferred host – the soft and juicy tomato – and is repelled by its alternative.

A plant can also use its sense of smell for protection. Some species talk to themselves, for a damaged leaf causes others nearby to get ready for attack – but not when the leaf is sealed in a plastic bag, as proof that an airborne signal is involved. Leaves chewed by insects pour out signals that are picked up by their neighbours, who prepare their own defences. Other species listen in to foreign messages. Tobacco seedlings grown close to a sagebrush bush switch on their own anti-insect chemicals when the bush is pruned. A few can even call for help. A beetle that chews a lima bean takes a real risk, for its victim sends out an aerial message that persuades leaves nearby to make a sugary secretion that attracts ants and wasps. The visitors then attack the beetle.

Plants can taste chemicals in solution, as well as smell them in the air. They extract information from the liquids that bathe their roots and flow across their leaves. Roots and shoots sense the presence of enemies and grow away from them. They hunt for food, too, for when a root hits a rich spot, it stops, sprouts and sucks up what is on offer. The substances that they pump into the soil may attract friends such as helpful fungi, but are also hijacked by enemies. Witchweeds are pests of tropical crops such as sugarcane. They grow on the roots of their host and – like the dodder – drain its vitality. They, too, pick up the taste of a dissolved substance used by the host to attract fungi. In the same way corn seedlings whose roots are chewed by grubs pump out a chemical that attracts predatory worms. The American black walnut scares away competitors with its own secretions and leaves a dead zone beneath its shade – and it is no coincidence that our own tongues are titillated by the poisons found in pepper, coffee, lettuce and more, which evolved not to satisfy the gourmet but to fight off an enemy.

Darwin himself saw that plants must have a sense of touch, for the climbers themselves, as soon as they contact a vertical object, change their behaviour, give up their wide sweeps and begin to twine. Roots, too, probe the soil and grow their way around a stone too large to move – although in this case they avoid, rather than embrace, the object. He found that the senses of touch and of direction interact, for a root held vertical will grow away from an object that blocks its path – but the same structure kept horizontal will always try to extend downwards, whatever obstacle is placed in the way.

Tree-huggers carry out what seems the entirely witless experiment of embracing a trunk to exchange energies with it; to inject their own vitality into the plant and to obtain some as yet undiscovered botanical spirit in return. In fact, to touch – or to hug – a plant has another unexpected effect, for it inhibits its growth. The pines of Highland forests are small, twisted and bent because they have been caressed – or battered – by the winds. Their equivalents around Down House live in calmer air and soar upwards. Identical seedlings grown in calm and windy places always end up with quite a different appearance. The plants are sensitive indeed

for to bend a young tomato plant for half a minute stops its growth
for a whole hour. That is why stormy places make for stunted trees.

The most conspicuous response to touch is that of the mimosa,
so admired by Shelley. Two centuries after his poem, botanists
tried a simple experiment: take a series of chemicals known to act
as hormones, dissolve them and water the mimosa to see which
bits of DNA respond. For every substance, a previously unknown
gene increased in activity by a hundred times. At first it looked as
if a crossroads in the hormone labyrinth had been discovered –
but sprinkling the plants with pure water had the same effect.
Mere contact had done the job. The first of the 'touch genes'
had been discovered. A drop of rain – or a gust of wind – causes
them to leap into action. More than five hundred separate parts
of DNA alter their activity when a leaf is prodded, some by ten
times and more. A hundred respond in the opposite way, but why,
nobody knows.

The touch genes do many things. They cause the cell wall
to firm up or loosen in response to stress and growth patterns to
alter as a result, which explains the gnarled Highland trees. Some
members of the clan respond to night and day, or to the seasons
as they pass (which is why leaves fall off in autumn), and others
are involved in disease resistance. They might some day be
engineered to give fruits that fall from the tree in a breeze as
soon as they are ripe or crops that grow tall in windy places. Half
the touch genes also respond when placed in darkness, but quite
why they do is still not clear. The sensitive plant is, it appears,
more sensitive than anyone imagined.

The inner world of plants has emerged as almost as rich as our
own. Darwin was cautious in his parallels between the sensory
and intellectual lives of the two kingdoms. The most he would
say is that 'It is impossible not to be struck with the resemblance
between the foregoing movements of plants and many of the actions
performed unconsciously by the lower animals.' Now we know that

the similarities go far further than he thought. One persuasive parallel involves the sense of touch. To everyone's surprise, some of the signal proteins used by plants to sense a gentle tap resemble certain molecules that do a similar job for us. They control our heartbeats, switch on hormones that determine growth and alter the blood chemicals that change mood from happy to depressed. In a further twist to the tale, young rats caressed by their mothers respond with an increase in the activity of certain genes related to those that react to touch in plants. A shortage of embraces stunts the animals' physical and emotional growth.

There is something magical in the way that scientific rationalism connects raindrops with heartbeats and battered trees with depressed infants. Shelley himself saw that science told us much that poetry cannot. He filled his Oxford room with electrical gadgetry and saw no contradiction between the worlds of the spirit and of science. He would have been delighted to learn that cooling passions are linked to falling leaves and that the Darwinian universal of shared ancestry shelters beneath its ample branches both the mimosa and its poet.

CHAPTER VII

A PERFECT FOWL

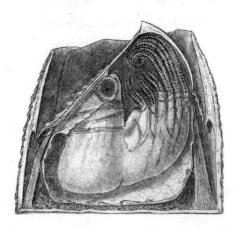

Sir Robert Moray was a spy for Cardinal Richelieu, a Freemason, a member of the Scottish army that took Newcastle from the English in 1640 and, in his spare time, the first President of the Royal Society. He wrote at length on the natural history of his native land and made a remarkable discovery, published in the Society's *Philosophical Transactions* in 1677. On a log on the shores of the island of Uist, he saw 'multitudes of little shells; having within them little birds perfectly shaped, supposed to be barnacles ... This bird ... I found so curiously and completely formed, that there appears nothing wanting, as to the external parts, for making up a perfect Sea-Fowl; ... the little bill like that of a goose, the eyes marked, the head, neck, breast, wings, tail and feet formed like those of other water fowl, to my best remembrance.' Sir Robert had the honesty to admit that he had never observed any of the adult animals but assured his readers that 'some credible persons have assured me that they have seen some as big as a fist'.

The myth of the shell-born birds, barnacle geese as we call them today, the shells themselves supposed to be the seeds of a certain tree, was already widespread. So embedded was the notion that for a time the barnacle goose was counted as a fish and could be eaten by Catholics on Fridays (Thomas Henry Huxley suggested that the mistake came about because such birds were common in Hibernia, or Ireland, and that the shift from *Hiberniculae* to *Barnaculae*, the term then used for barnacles, was easy enough).

The idea of a bird-bearing tree is foolish, but it arises from an ancient and accurate observation – that the adult form of many creatures is quite distinct from that of their eggs or embryos. The untrained eye finds it hard to tell juveniles apart. A month-old human foetus is almost identical to that of a chimpanzee, the inside of a goose egg looks much like that of an ostrich and a barnacle larva is not very different from those of its relatives among the lobsters and crabs. Even the founder of modern embryology, Karl von Baer, found it difficult. In 1828, he wrote that 'I have two embryos preserved in alcohol that I forgot to label. At present I am unable to determine the genus to which they belong. They may be lizards, small birds, or even mammals.'

The word 'evolution' (which does not appear in *The Origin of Species*) was first applied to the unfolding of the body as egg is transformed into adult. Development is the imposition of pattern upon a formless mass. Most animals, from barnacles to geese, share the same basic types of cells. As the embryo grows they are organised to make a crab, a goose or an ostrich, a man or a bat. That grand reshuffling builds new and complicated body shapes from the same raw material. As it does it hides the logic upon which bodies are built. Adult anatomy makes much more sense when seen through the eyes of the embryo and *The Origin* itself used the similarity of the juvenile stages of apparently unrelated beings to argue that 'community of embryonic structure reveals community of descent'.

Its author saw that many creatures showed 'unity of type', a deep similarity manifest in the young but largely hidden by the complexity of the adult form. Many embryos – those of barnacles included – consisted of repeated segments that are multiplied, reduced or rearranged to produce an adult. An increase or decrease in number or a shift in pattern of growth can generate a vast diversity of size and shape. Evolution, Darwin realised, works as much by the manipulation of repeated units as by tinkering with the details of individual organs as they grow.

The idea finds new life in modern biology, which reveals affinities among the embryos of even distant creatures. DNA, like the bodies it builds, is itself based on a series of variations on a structural theme. As egg becomes adult, complex organs – eyes, ears, hands and brains – are pieced together from elements that can clearly be distinguished only in the embryo.

Nowhere is the contrast between young and old more remarkable than among the barnacles. Once, such creatures were said to be snails because of their solid shells (and a well-known professor of zoology – or of biochemistry masquerading under that title – once tried to convince me that they do belong to that family). In fact, they are jointed-limb animals not unlike crabs, spiders or flies. Their ancestors lived free in the oceans but now many spend most of their lives in a prison cell. Barnacles are close kin not to limpets as once imagined, but to shrimps and lobsters. That affinity was discovered in the 1820s by an army surgeon based in Ireland but for many years the group – the cirripedes or 'curly-footed' to give them their technical name – seemed no more than obscure. Few biologists could be bothered with such tedious creatures.

Until, that is, Charles Darwin spent a sixth of his scientific career on them. His eight years of research, in the interval between the *Beagle* voyage and *The Origin of Species*, showed that

animals, dull as they might appear, had lessons not just for naturalists but for biology as a whole. As he came slowly to the idea
that life was not fixed but might change, he was warned that 'no
one has the right to examine the question of species who has
not minutely described many'. Perfectionist as ever, he agreed:
he realised that to understand the logic of life he needed to
become an expert on a single group. Today's biologists are
obsessed with 'model organisms' – fruit flies, a certain worm,
mice, mustard plants and even humans – that might, when their
secrets are unveiled, be exemplars of evolution on a wider stage.
Cash pours in, and optimists hope that to understand their
favourites in detail will illuminate the science of life. Some of
the supposed archetypes turn out, alas, to be quite untypical
even of the group to which they belong (and the fruit fly itself
falls into that category). The first model organisms of all were
Darwin's barnacles. He made – by luck or judgement – an
excellent choice.

He wrote four books – well over a thousand pages – on their
taxonomy, their embryos and their fossils. Some species were
bizarre; so distinct from the familiar rock-dwellers of Welsh or
Scottish shores that any kinship appeared almost as improbable as
did an affinity to geese. The young naturalist did his job so well
that, at the age of forty-four, he was given the Royal Society's Gold
Medal for his work. His solid volumes remain a standard reference
work today. More important, they laid the foundation of a central
theme of evolution: that the embryo is the key to the adult.

Darwin's attention was drawn to cirripedes when, as a medical
student in Edinburgh, he spent weeks in the hunt for marine
animals in the Firth of Forth. There he fell under the influence
of the zoologist Robert Grant, who introduced him to life on the
sea shore and encouraged him to publish his first scientific paper
(Grant later became Professor of Comparative Anatomy at University
College London, but the two fell out over the issue of whether

animals showed inevitable progress from low to high and almost never spoke again even when they worked in the same street). Almost a decade after his studies on the chilly shores of the Forth, the *Beagle*'s naturalist found on the shores of the Chonos archipelago off the coast of Chile an enigmatic soft-bodied creature 2.5 millimetres long drilling into a conch shell. At first he thought it was a worm, but under the lens it became clear that the creature was a great anomaly, for naked as it might be, it looked very like a British barnacle. Could the animal, in spite of its lack of a shell, be related to the denizens of a Scottish shore? If so, how – and why was the creature so different?

Darwin tried to find out. He planned at first just to sort out the Chilean creature's place in nature, but as the work went on, he found more and more distinct and – on the face of it – aberrant kinds. Soon he began to notice what appeared to be intermediate forms between them and series that showed greater or lesser affinity to each other. Oppressed as he was by the tedium of the task ('I may as well do it, as any one else'), barnacles sharpened in his mind the idea – already implanted, as his notebooks show – that one species might change into another. Perhaps, he became convinced, all barnacles – all animals – descended from a common ancestor that could be tracked further and further into the past. Five years after his cirripede opus, that radical notion became the theme of *The Origin of Species*.

The juvenile stages revealed unexpected connections between the South American borer and its Scottish kin. That lesson, learned on the shores of Chile, has grown into the science of evolutionary developmental biology, which unites barnacles from across the world with each other, with crabs and lobsters and even with geese. It reveals the common foundations upon which all animals are built.

In the first days of development, many creatures resemble one another more than they do when they become adults for each shares a series of genes that lay down the basic body plan, from head to tail. Such genes are control switches in the journey from fertilisation to the grave. They shepherd the egg towards adulthood. Errors lead to dramatic shifts in form – eyes transformed to legs in fruit flies, lambs with two heads or extra fingers in human babies – together with more persistent changes such as those that made birds from dinosaurs or barnacles from the ancestors of crabs.

Darwin was sent specimens from across the globe. Some would, he realised, stretch the belief of his fellows and he wrote to a colleague about his discoveries that 'You will think me a Baron Münchausen among naturalists.' His first job was to describe what the animals looked like. As ever, he told a simple story in plain prose.

His introductory paragraph is a sober account of what most people imagine such creatures to be: 'Almost every one who has walked over a rocky shore knows that a barnacle or acorn-shell is an irregular cone, formed generally of six compartments, with an orifice at the top, closed by a neatly-fitted, moveable lid, or operculum. Within this shell the animal's body is lodged; and through a slit in the lid, it has the power of protruding six pairs of articulated cirri or legs, and of securing by their means any prey brought by the waters within their reach. The basis is firmly cemented to the surface of attachment.'

That statement introduced the immense variety of cirripede lives. More than twelve hundred different kinds are known and no doubt many more remain to be discovered. All live in salt water. They fall into two main groups, those with a stalk (the goose barnacles, named in homage to the ideas of Sir Robert Moray, and a delicacy in many parts of the world) and those without, many of them, like the familiar acorn barnacle, attached to rocks and other marine structures. All, or almost all, have jointed legs, often tucked away within a shell. Many use them as a net to sweep the seas, while the stalked versions depend more on the movements of the water to bring food. Unlike their relatives the crabs and lobsters, barnacles do not moult their skeletons to grow.

Instead their plates increase in size as the animal gets older. Some species sit on rocks, while other kinds burrow through solid stone or into snail shells or spend most of their time afloat. Yet more are parasites of crabs, jellyfish and starfish. Some among that group are so specialised that, when adult, they look more like a fungus than an animal.

Like insects, barnacles have a head and thorax and, in a few species, what might be the remains of an abdomen. Like them, they have six pairs of jointed legs, fewer than the prawns and lobsters, who have ten. Each leg is covered with hairs and together they lash the sea. The familiar shore versions spend their lives upside down for they stand on their heads and wave their feet in the water.

Those found on rocky shores live in a fortress made of around six tough plates, based, like a snail shell, on a limestone-like substance. Different varieties have more or fewer segments of body armour and many pages of Darwin's four books on the creatures are devoted to the minutiae of how their plates might sort out their patterns of relationship. For the common British form, an additional two plates act as a lid, which opens to let out the legs at high tide and closes to keep in water when the creatures are exposed to the air (which for some individuals means all the time except for a few days each month at spring tide). The mouth has structures that chew and grind and look a little like those of crabs and even of cockroaches. Some species excrete through their mouths as their anus has faded away. Tucked away in the dark, the adult barnacles lose their eyes. The nervous system, too, is reduced when compared with that of their free-living relatives.

Dull as a cloistered existence within a gloomy fortress might be, all barnacles have a remarkable sex life. Like all good biologists, Darwin spent a lot of time on that topic. He found a wild diversity of reproductive habit. The textbooks of his day said that all known species were hermaphrodites but many, he found, were not. Some have two sexes, some are male when young and female later, and some are true hermaphrodites – while a few among that group secrete small males around their bisexual persons in case they might be useful. Many of those with two sexes spend their adult lives fixed to a single spot. As a result every male must constantly wave his penis, erected at the cost of lots of hydraulic energy, to reach out and tap his neighbours in the hope that at least one might be a female. Those who find themselves in a sparse and scattered group must, if they are to succeed, grow a longer organ than those who live in a crowd. A female, once tapped, may copulate with half a dozen males in series and then pump out most of their seminal fluid as not up to scratch. Her fertilised eggs soon develop into the first of several larval stages.

The young biologist's studies on cirripede sex brought forth some poetic paragraphs. The male organ of a certain species was 'wonderfully developed . . . it must equal between eight and nine times the entire length of the animal! . . . there [it] lies coiled up, like a great worm . . . there is no mouth, no stomach, no thorax, no abdomen, and no appendages or limbs of any kind'. In another the males were reduced to parasites within the female: 'thus fixed & half embedded in the flesh of their wives they pass their whole lives & can never move again'. The creatures hold the record for relative penis size, while a certain beetle comes next, at twice its body length. The organ comes at a price, for individuals from wave-battered shores have shorter and stouter members than do those from calmer places as the male finds it hard to control a lengthy structure in a turbulent world. So expensive are such massive genitals that many males lose them at the end of each season and grow a new set the following year.

Barnacles are remarkable for reasons that stretch beyond the penis. They stick to the rock with a sophisticated cement, a protein that repels water. The stuff is the toughest known natural glue. Like an epoxy adhesive, the material is secreted as a clear fluid with two components. When they mix, cross-links are made between the molecules and its manufacturer becomes almost impossible to dislodge. So powerful is the bond that some of the substances involved may soon be used in surgery.

The creatures cling to ships just as avidly as they do to rocks. Charles Darwin, exhausted by his years of work on them, once wrote that 'I hate a Barnacle as no man ever did before, not even a sailor in a slow-sailing ship', and mariners had good reason to despise the animals. The *Beagle* herself had to have her bottom cleaned several times in the cruise around South America. A ship uses 40 per cent more fuel when covered with marine organisms than when its surface is smooth − which is expensive and, in these days of the greenhouse effect, also to be deplored on ecological grounds. Poisonous paints were once used to keep the bottom clean, but as many caused sea snails to change sex most of them have been banned. The best protection is to find a finish to which the animals cannot attach. Given that they can adhere to a non-stick saucepan, the job is not simple, although paints with added carbon nanotubes offer some hope. Some corals and seaweeds manage to stay free of such creatures not with poisons but with chemicals that scare them off. Those unknown substances will make the fortune of the first scientist to extract them.

Barnacles have been passengers since long before ships sailed the oceans. Many creatures suffer their attentions. Humpback and grey whales bear large white patches of thousands, some of the individuals several centimetres across. The larvae pick up the scent of their host as they float through the sea and move towards it. Then they dig into the skin − and the huge beasts pay a price in energy as they drag their hangers-on through the sea. The whale retaliates with skin grown at a rate three hundred times that of our own in an attempt to slough off its passengers. Some marine mammals secrete enzymes that dissolve the glue and help keep their foes at bay, while grey whales come in to land to try to scrape the hitch-hikers off. Dolphins move fast and are safe from such visitors, who are washed off before they can fix on, but big sharks, who idle through the water, are also free of the pests. Shark skin is covered with tiny ridges – and a film has been developed that mimics its structure. It may find a place in the world of commerce.

Other species of barnacle hitch lifts on the gills of fish, or live around the deep-sea vents that belch out hot rich water that nourishes a thick soup, just right for a filter-feeder. Not all cirripedes have a settled way of life, for some float blithely through the seas and never touch a solid object while yet others bore into coral reefs.

The most aberrant kinds take up a sinister profession. From a whale's or a matelot's point of view, a barnacle is an irritant, but little more. For crabs faced with a marauding cirripede, the situation is far worse. A certain group live as parasites within their living bodies. Their macabre habits give an insight into the spectacular diversity of form that evolution can come up with when it generates variations upon a body plan.

First, a female larva lands on its victim and finds a soft spot in the creature's armour. Then she stabs it with a hollow needle and fires a few of her own cells through. She dies at once. The baleful blob finds its way to the lower part of the crab's body and sends out fine tendrils that run through the host's entire anatomy. They grow to make a mesh that looks more like a mould than a marine animal, and suck in food. The crab stays healthy and continues to eat as fast as it can to feed the visitor. When the time is ripe, the parasite opens up a small hole to the outside and awaits the arrival of a mate – a male larva. Should a male appear it inserts its spiny self through the hole and seals it up to prevent the entry of a rival. Now the crab is in real trouble. The male parasite fertilises his partner and she begins to pump out thousands of larvae. The victim's whole economy is hijacked and it can no longer grow, shed its skin or even replace a damaged part. Instead, sometimes for years, it devotes its energies to its inner barnacle.

Soon the crab, male or female, is spayed by the unwelcome visitor. A castrated male crab starts to look, and behave, just like a female. Both sexes now act as mothers – but mothers who care for another's interests. They develop a pouch on the underside that resembles that made by a healthy female just before she releases her offspring and spreads them through the water with sweeps of her claws. The unfortunate crabs again wave a mass of newborns on their way, but now they are not their own progeny but barnacle larvae ready for the next target.

The diversity of cirripede lives confused Victorian biologists, who could see no logic in their variety of shape and habit. Large parts of Darwin's work turned, in the traditions of the time, on an attempt to understand how the various species are related to each other and to find where the group as a whole fits into the animal world. That pastime – taxonomy, as the science is called – was once little more than stamp-collecting, but for him it became the raw material for a deep insight into biology. His classification was based in part upon the solid plates that surround most settled barnacles and he persuaded himself that he could see a hint of order that reflected their ancient ties (even if he stayed confused by his Chilean burrower and by the parasites). The scheme has been much modified and a system based on the pattern of adult plates remains ambiguous.

The new philately – molecular genetics – studies shared descent in DNA itself, rather than in what it makes. The confident assumption that mutations accumulate at a regular rate to give steady divergence generates a family tree of the group's evolution. Well-dated fossils can, in principle, be used to measure how fast the changes happen, to give a molecular clock. The tree suggests that stalked forms came first and the others followed on. The double helix also hints that the protective plates emerged after a jointed-legged animal had taken the decision to settle down and wait for food to arrive rather than going out to hunt for it. Several

of the classical groups, such as the naked barnacles that so confused Darwin, appear to be a mixed bunch with distinct origins,
and even the rock-dwellers are an assorted lot. The deep-sea vent
types and the parasites are each, in contrast, a group of true relatives. The genes also confirm his view that the cirripedes as a
whole fall into the larger family of crabs and lobsters – the
Crustacea – and into the wider clan of insects, spiders and other
jointed-legged animals. Some claim, on the basis of their shared
molecules, that insects themselves are no more than a specialised
group of crustaceans that reached the land. If so, they reveal an
unexpected unity between barnacles and butterflies.

Whatever the details of their family connections, the diversity
of cirripede life began long ago. Two of the great evolutionist's
books deal with their fossils. They are not, perhaps, the most riv-
eting of his works but they make, nevertheless, a forceful case that
today's kinds descend from forms now long extinct. Darwin
referred to modern times as the 'Age of Barnacles', and at least in
terms of the number of species known he was right, for their fossils are not abundant and can be hard to identify because the plates
fall apart after death. Cirripedes do not appear in the rocks in any
numbers until the demise of the dinosaurs, sixty-five million years
before the present. A few spots do reveal good evidence of their
passage. The Red Crag deposits of East Anglia were laid down in
the cool Essex seas of two million years ago. The rust-coloured
rocks are still full of their protective plates, mixed in with snail
shells and the teeth of the largest sharks ever to have lived. Further
from home, impressive strata in the south of Spain record, in the
mix of their cirripede species, the rise and fall of a vanished sea.
Their petrified memorials also show that whale barnacles have
been around for at least two million years, for a bed in Ecuador is
filled with their remains as a hint that the whales once bred there,
as they still do, just off the nation's coast. A 164,000-year-old
whale barnacle specimen from a human settlement in an African
cave shows that our ancestors have long eaten those huge marine

mammals. They scraped off the external parasites and may have cooked them.

No more than a few very ancient specimens have been found. A fossil from three hundred million years ago looks rather like a modern barnacle. Another well-preserved remnant, found in Herefordshire, from a hundred million years earlier – about the time of the first land animals – resembles the larva of a modern cirripede and hints that the group was well into its stride by then. The low-growing forms found on rocks, abundant as they now are, emerged far later, perhaps no more than a hundred and forty million years before the present, when *Archaeopteryx* walked the Earth.

Anatomy, genes and fossils each place the barnacles in close association with crabs and lobsters and in less intimate kinship with insects, spiders and more. That larger group makes its presence obvious early in the record, in the Cambrian, more than half a billion years before the present, the era in which life first left abundant evidence of its passing. Some of the mysterious creatures with bizarre body plans found just before that time and once claimed to represent a unique and vanished fauna may in fact have been crustaceans. A molecular clock of the whole group puts the origin of the barnacle lineage well back into the Cambrian, or perhaps earlier, even if no earlier remains have yet been found. If the clock can be trusted, the first cirripedes may have emerged as part of the vast outburst of diversity among jointed-legged animals from lobsters to insects, which began then and is still evident today.

What sparked off the barnacle big bang? Why did they, like their crab and insect brethren, evolve into such diversity of form? And why did vertebrates, the group to which we and the barnacle goose belong, do the same many millions of years later? Backboned animals are less diverse in their body form than are cirripedes, but they include creatures as different as mackerel, toads, pythons and vultures. Why was their evolution, like that of barnacles, so radical while groups such as sponges or flatworms remained, in comparison, tediously conservative? The answer began to emerge from Darwin's labours over the Down House microscope.

Its owner was the first to identify a barnacle larva, from his strange shell-borer from Chile. As he dissected more and more species and examined their juvenile forms a great truth began to dawn: that the creatures were far more distinct from each other as adults than they were in their early stages. From Scottish rock-dweller to naked Chilean and from tasty marine snack to the sinister enemy of crabs, the juvenile forms of the various species were very similar. Even better, they looked quite like the equivalent phases in crabs and lobsters. Darwin's excitement at this discovery is manifest: he writes of a larva 'with six pairs of beautifully constructed natatory legs, a pair of magnificent compound eyes, and extremely complex antennæ'. He knew that he had hit upon a crucial piece of evidence for evolution (although his children laughed because the sentence read like a newspaper advertisement by a cirripede manufacturer).

Most barnacles release thousands of tiny fertilised eggs into the sea. Each goes through a series of stages, in most cases as a form that floats free in the plankton. The first has jointed limbs attached to a soft and flattened body. The young animal has an eye spot, sensitive even to dim light, that allows it to choose the level at which it floats. Soon it develops jaws and antennae and starts to feed. It goes through several moults and in time becomes a strong-swimming form with a tough outer coat. Those mature larvae prefer to stay near the surface, do not eat and can be carried far from where they were born. They must find a place to settle down, or – as almost all do – they will die. Some stumble upon a rock, or a whale, or a crab, and glue themselves on with their antennae. The rock- or whale-dwelling species put out a chemical message – a protein hormone – that invites others to join the colony. For them, every visitor is welcome, for a male must land within penis-length of a female if he is to have a chance to pass on his genes and the more there are the better.

Much as the first stages of many species might resemble each other as they float through the seas, some – like those that amused the Down House children – do have aberrant juveniles, adapted to their own special way of life. Those of the burrowers cannot swim but scuttle about on the bottom using their antennae as feet. Crab parasites have abandoned the first few stages altogether and hatch as jawed and hungry forms that search for new victims at once. Natural selection is at work on the larval stages, which have to adapt themselves to nature's challenges just as grown-ups do. Even so, the young reveal far more about the group's internal affinities than do the much-modified adults. They show how cirripedes and their relatives are based on a theme with variations.

The same is true of the embryo on a wider stage. That of a barnacle goose is almost identical to the contents of a vulture egg and an embryonic human looks rather like that of a mouse or, indeed, if looked at early enough, of a goose. What emerges into the world is quite distinct from what can be seen as development begins. Now we understand why.

Adult cirripedes apart from the crab parasites are – like lobsters and insects – arranged in obvious sections, with a head and a thorax divided into six segments, but they lack an abdomen, found in almost all their relatives. We do not often think of ourselves as segmented creatures, but the vertebrate body is, like that of a barnacle or a lobster, also based on a series of distinct units, arranged from front to back. The human head, thorax and abdomen are obvious enough but our muscles, or our brain-case, show little sign of order. A glance at the embryo, however, reveals that men and women, like their submarine relatives, are constructed from a series of modules, neatly arranged in early life but shuffled around and modified as growth proceeds.

The remains of our watery past as primitive fish, together with the juvenile forms of our relatives among fish, snakes and birds say more. They show how the building blocks have multiplied and rearranged themselves to make the complicated creatures of today.

Just three of the thirty or so major divisions of the animal world are organised in obvious segments; they include the worms, the jointed-legged creatures such as insects, spiders, lobsters and barnacles, and the animals with backbones. For all of them a subdivided way of life has been an evolutionary triumph.

Segmented beings make their first appearance at – or even before – the first signs of the fossil record. They played a large part in the Cambrian explosion of diversity. Fossils from that time show how the addition of new pieces to a simple body, like beads on a string, can spark off a burst of change. Many of its strange animals were worm-like beasts, or had jointed legs and external skeletons. In time they added more and more sections. As they did, they evolved into a wild diversity of form. One ancient marine group, the trilobites (now extinct), started off with around eight segments. In time, some kinds ended up with a hundred and others with three. That process then, for some reason, reversed itself and at the peak of their success most trilobites had at most thirty-five separate elements.

As Darwin noticed, barnacles and their relatives have been through the same process of increase, decrease and divergence. He persuaded himself that the archetypal crustacean, the ancestor of both cirripedes and lobsters, was based on twenty-one parts, divided among head, middle and abdomen. Many modern species have six elements in the head, six in the thorax (the middle part of the body) and five in the last, abdominal, section. Some have multiplied and modified particular elements while others have done the opposite. Lobsters, for example, have many more paired and jointed appendages – legs and swimmerets plus others used to mate or to help brood the young – than do crabs, while the barnacles themselves lack the whole rear segment of the body. They are the Manx cats of the crustacean world and, for that matter, are an excellent analogue of the first birds, which were dinosaurs who shook off their tails.

Goethe – philosopher, scientist and author of *Faust* – had, well before the *Beagle* voyage, noticed hints of pattern within the bodies of fish, birds and mammals. He came up with a universal theory of anatomy, based on the notion that vertebrae – the individual sections of the backbone – were units from which many of our various parts were derived. The leaf, he imagined, had the same role in plants. Goethe saw life as emerging from a sort of biological Proteus; a simple component that could be multiplied and modified into a diversity of structures, the skull most of all. He was wrong in the details, but his idea contains an element of truth.

Although the simplistic claim, never made by Darwin, that animals relive their ancient history as they develop from the egg is wrong, the embryo is a reminder of where we came from. The shift from fertilised egg – a formless ball of protoplasm – to man or woman looks complex but is in its basics simple. As in origami, a limited set of instructions persuades pattern to emerge from simplicity. As the embryo folds itself into being, its past unfolds before our eyes.

Hints of order soon appear. A fertilised egg divides to form a ball of cells, which in time turns itself inside out and becomes attached to the wall of the uterus. It lengthens, and a ridge – which soon becomes a tube, the precursor of the spinal cord and brain – forms along the upper surface. The masses of tissue on either side then begin to break up into a series of evenly spaced blocks called somites. Those near the front appear first, and tissue stains show that ordered structures arranged from front to back are present long before the somites themselves become visible.

The somites in their rows look simple, but they give rise to complex structures, some of which have no obvious hint of regularity; to vertebrae (which would have pleased Goethe), to ribs, to muscles of the back and the limbs, to skin and tendons and even to certain blood vessels. The organised nature of vertebrae is obvious enough, but to the untutored eye the muscles of the leg or the skin on the back give no hint of segmentation. Even so, they – like many other organs – began as blocks of tissue.

As development goes on, the front half of one somite fuses with the back of the somite ahead of it to form the precursors of vertebrae – the repeated units of the spine, the structure shared by fish, frogs, snakes, birds and humans. They surround the spinal cord with a protective and flexible sheath that solidifies as bone is formed. The process is controlled by special growth factors, which sometimes go wrong. That has an echo of Goethe, for after a failed attempt by the East Germans in the 1960s to conserve his corpse his body was stripped of flesh – and it was revealed that the great poet suffered from a debilitating fusion of several spinal bones.

How can a uniform embryonic tissue break up into segments and then into distinct organs? In 1891, William Bateson – later the rediscoverer of Gregor Mendel's work – came up with a 'vibratory theory of the repetition of parts': the notion that a flow of chemicals did the job. Just as waves on the sea create ripples on the sand, their equivalents in the body stamp order on to disorder. A century and more later, he was proved right.

As the embryo develops, chemical signals that promote growth diffuse from its rear end towards the front. They are matched by a second molecular message that travels in the opposite direction and tells the tissue to mature and stop dividing. Each potential somite has an internal timer that instructs genes to work for the appropriate time and then to switch off. When the signal arrives, the clock starts. The somites each contain a hundred or more genes that cycle in and out of phase with each other, many with opposed effects on cell division, growth and movement. Together they build the block of tissue – and the genes that do the job are similar in mice, chickens and barnacles, proof that the basic rules of segmentation began before they last shared an ancestor, long ago.

Vertebrae still retain strong hints of their segmented history. Their numbers vary from species to species. Most people have thirty-three of the bones (with several fused together), geese have more (particularly in their necks), but snakes may have over five hundred. The vast increase among the serpents arises because the clock within each of their somites ticks several times faster than does our own. As a result, the mass of tissue is converted into many more segments in the time available – and the animal gains its long and flexible backbone. Perhaps the same is true in the goose's neck.

Each human vertebra has a personality of its own. Some are reduced to form a vestigial tail and others fuse to make a solid block at the lower end. Those in the upper back grow large spines to which muscles are attached while the seven vertebrae in the neck are specialised to allow the head to move from side to side or up and down. Whatever its task, every vertebra has, as a reminder of its shared embryonic experience, a strong resemblance to its neighbours.

The skull, or so it seems, is different. Its twenty-two bones show no obvious signs of segmentation and, apart from the lower jaw, all are fused together. The cranium is a round case with many openings and a variety of special structures such as the eye-socket, the teeth, the jaws and the ear. It appears at first sight to have little in common with the backbone upon which it perches. Now, science has shown that – as Goethe had hoped – it does.

Once again, the embryo is the key. The skull is in part built from somites (with most of the rest formed from bone laid down by precursors of other tissues). The genes prove that parts of at least the first two somites contribute to the skull. As further evidence, mouse mutations that damage the somite signalling machinery also affect the cranium. The skull, complicated as it might be, began as just another block in the body's support axis.

Its anatomy, its fossils and its genes say a lot about the way in which segmentation can make complex structures from simple precursors. The organs of sense and of thought that live within the skull have long been used by anti-evolutionists to cast doubt on Darwinism. In fact, every part of the skull puts paid to the 'argument from design', the ancient and threadbare claim that complex organs must need a designer. Darwin himself quoted the eye as evidence against that notion. The ear makes the case even better and has the additional advantage that fossils can join the embryos to show how evolution has cobbled together solutions from whatever is available. If a designer did the same, he would lose his job.

The human ear has an outer, middle and inner section. Together they pick up vibrations from the outside world. The outer ear receives the sound waves, the middle amplifies them with the help of physical movements of a set of bony levers while the inner ear transforms that mechanical energy into pulses of liquid and, in the final stage, into electrical and chemical impulses that pass to the brain. The inner ear also gives its owner a sense of physical position and of acceleration or deceleration.

The organ in its intricacy is witness to the power of variation on a theme and to the joys of improvisation. Genes, embryos and fossils combine to show that it evolved from the skeletons of ancient fish – and that the human ear shares some of its components even with the sense organs of barnacles.

All land vertebrates have some form of ear. The outer part of the organ, the pinna – that elegant appendage on each side of the head – is rather new, for frogs, reptiles and birds do not bother with it. It is made from the same cartilage and skin as much of the rest of the body surface. Darwin noted that in humans and apes, unlike dogs, it could not move, perhaps because as large animals able to climb trees they no longer needed eternal vigilance. He was told by 'the celebrated sculptor, Mr Woolner' that, while working on the figure of Puck, he had noted that some people had a small point folded in from the outer margin –

perhaps, Darwin suggested, a vestige of a pointed ear. The structure is now known as Darwin's point.

The membrane of the eardrum is the gateway to the middle ear. It vibrates when sound strikes and passes the energy to three tiny bones – the hammer, the anvil and the stirrup, each named after their shape – that act as levers. Each fits into the next and together they amplify the movements of the drum into larger movements that are passed down the chain to a small, membrane-covered window on the surface of the liquid-filled inner ear. Because the eardrum itself is larger than the tiny window, the system increases the pressure upon it by twenty times or more. As a safety measure, tiny muscles attached to two of the bones damp down the harmful effects of loud noise. The inner ear transforms the energy of sound waves into messages about intensity, pitch and direction that pass to the brain.

The three-part middle ear is as specific to the mammals as a whole as is hair or milk, for all other land vertebrates have just a single bone in that organ. The structure is a wonderful example of how repeated elements can be used for a diversity of ends. Fossils, embryos and the animals of today paint a remarkable picture of how a body built on modules adapts itself when faced with a new challenge.

The ability to pick up waves in air or water began long ago. Most marine creatures can do it, with the help of simple sensory cells. Fish, too, are quite modest in their talents. Living as they do in water, an excellent conductor of wave energy, they manage with just a series of pressure sensors on either side of the head and body. Land animals face a more demanding task, for they need to amplify the feeble power of waves in air. They use the middle ear to do so. Fossils show that the structure appeared around two hundred and fifty million years ago, a hundred million years after the direct ancestors of mammals split off from their reptile ancestors, and seventy-five million after the bird and lizard lineages diverged from the same source. All three lineages developed a bony lever independently and with its help each of them improved its own ability to detect high-pitched sounds. Reptile and bird eardrums still connect to the inner ear via just a single bone, the stirrup. A set of triple levers, unique as it is to mammals, does a lot to improve our own hearing. Each of the bones can be traced to simpler structures in primitive animals with an easier way of life.

Anatomists came up with the first evidence of the history of the ear at about the time that Darwin began his barnacle work. They saw that in the first phases of development the embryos of fish, reptiles and mammals generate, just after the somites appear, a series of looped arches on either side of the front end. Those six repeated structures grow into matched pouches on each side of the developing head.

Four hundred million years ago, the only animals with backbones were flat-headed fish that swam in an immemorial sea. Their bodies were covered in bony plates and those primeval vertebrates ate without the benefit of jaws. In time they were succeeded by fish-like creatures with necks and simple limbs. They clambered ashore around 365 million years ago and evolved into frogs, lizards, birds and people. The fossils of those antecedents of all the vertebrates tell the story of the middle ear. It confirms that told by the embryo and by the genes.

In antediluvian fish, the arches were supports for the gills, the structures that extract oxygen from the water. They did a simple job that lasted for millions of years. As their descendants grew bolder and moved on to land, natural selection spotted the opportunity offered by a repeated structure. In time the arch nearest the front was hijacked to become modified into the first jaws of all. The lower and the upper jaw of all vertebrates, one hinged into the other, hence trace their origin to an ancient aid to fish respiration. The second arch was then picked up to make a bone that connects the upper jaw to the brain-case. As their descendants crawled on to land, that structure evolved into a lever able to amplify sound.

Lizards and their descendants the birds had but a single such bone. Then, as the immediate ancestors of modern mammals appeared, the ear began to commandeer other parts of its ancestors' anatomy. First, the position of the hinge between the upper and lower jaw shifted compared with that found in reptiles. As it did, it freed a bone within the upper jaw, and another one within the lower. Those redundant structures were seized by evolution to make the hammer and anvil bones of the middle ear – which means that we hear, in part, with what our ancestors chewed with. Fossils of the first mammals as they began to evolve from their reptilian ancestors three hundred million years ago reveal the whole process, in all its steps, in a series of creatures with more and more complete middle ears. The shift from food processor to hearing aid happened several times in different mammal lineages, most of which are now extinct. Those small creatures of our first days ate insects and moved around at night – and any improvement in the ability to hear would have been useful indeed. Anatomy agrees about the ear, for the nerves which serve the stirrup bone branch from that to the face, while those to the other two are offshoots of a different nerve (a fact otherwise inexplicable).

Each of the three bones of the middle ear hence comes via a different route from two of the fish gill arches. Those ancient structures have also been taken up for other ends. In mammals remnants of the first arch help make some of the chewing muscles. The second evolved into some of the muscles of the face and into the bone in the neck that supports the tongue and is important in speech.

The embryo tells the same story, for as it develops the famous arches can be seen to reinvent themselves to become parts of the middle ear. The genes that build them, too, resemble others still active in the gill-slits of modern fish. The case for the middle ear as a pastiche based on an ancient marine structure is watertight.

The inner ear, deep within the skull, is another legacy of an extinct fish – and even of an early barnacle. It, too, reveals its history in fossils, embryos and DNA. The sea is a noisy place, for water is almost transparent to sound. Whales sing, fish grunt and crustaceans join in; the pistol shrimp gains its name from the loud clicks it makes with its claws, while its relative the mantis shrimp, whose claw can break a fisherman's finger, emits a deep rumble that frightens off predators. Lobsters, in the same way, make alarm signals by scraping their antennae across a ridged section of carapace. The larvae of lobsters and crabs – with their close resemblance to those of barnacles – pick up the roar made by waves upon a reef, and make their way towards the sound from kilometres away. Fish are even more responsive to such stimuli.

All three groups use the same fundamental mechanism for those jobs: a set of specialised pressure-sensitive cells filled with jelly, into which is affixed a hair-like structure that extends to the outside. A wave – caused by a current, the echoes of a surf-battered shore or the movements of a nearby enemy or friend – causes the hair to flex and the cell to pick up that movement, to transform it into chemical and electrical activity and transmit the information to the brain. Our own inner ear has just the same arrangement, for the physical movements of the middle ear bones make waves that disturb a set of sensitive hairs, which in turn generate a nervous impulse. Damage to a certain gene causes deafness and a search through fish DNA finds the same gene active in the pressure-sensitive cells. So similar are the two systems that fish are used to test drugs that might damage hearing if used on ourselves.

As Wagnerians can attest, human ears do rather more than just notice changes in pressure. Our ability to tell notes apart, impressive as it might be, emerges – once more – from expansion and diversification, in this case of the simple fish system into a series of sensory cells with different sensitivities to particular tones, multiplied and arranged in order within a long coiled structure. The reptile version is short, which means that snakes and their allies can hear only low sounds, that of birds intermediate, and the mammal inner ear sensor the longest of all. The story of the ear is of make do and mend, and of multiplied structures modified by natural selection for a new and different end. Perfect pitch, for those who have it, has been reached by most imperfect means.

The double helix shows that the modular plan upon which life as a whole is built goes back to long before the evolution of barnacles, geese or men. Whole sections of the molecule have been multiplied or lost as evolution made crustacean, birds or mammals.

Certain fruit-fly mutations that double up the number of wings or antennae are due to changes in the genes that control the passage from embryo to adult. Such homeobox genes, as they are called after a short repeated DNA-binding sequence (or 'box') found in all of them, alter the timing and rate of growth of various segments of the genetic material and change the shape of what they build. They are a molecular mirror of Darwin's discoveries among the barnacles: of duplication, reshuffling and deletions of parts. As they multiply, such sections diverge to take up new tasks and on the way remove another plank of the creationist cause: that evolution can only remove information and cannot create it.

One surprise in modern genetics was to find how small the molecular divergence among animals actually is. A goose and a chicken are almost identical at the DNA level and neither is particularly distinct from a human. The barnacles, in turn, are close to the crabs and not very different from flies. Many of their genes have changed not at all in the millions of years since they diverged. Geese and barnacles may look quite unalike – just as cars and aeroplanes are distinct even if each is built from the same basic elements. Genes make the nuts and bolts of the body. What they make is put together in different combinations and instructed when and where to do their job. Some act as switches that activate or suppress the activity of particular genes in the embryo. The evolution of segmented animals depends in large part on their gain or loss. In some creatures they are arranged in the same order as the body parts, with head first, then the middle section and then the abdomen, but that neat arrangement is often disrupted as the homeoboxes are broken up into separate clusters or scrambled altogether. Different creatures have from around four homeoboxes to four dozen or so. Their presence in barnacles and buzzards, sea-urchins and squirrels, or spiders and snails, suggests that the universal ancestor of all those animals was a segmented worm-like creature in an ancient sea, with around eight of the famous genes.

Our own homeobox system is arranged in four clusters with about ten members in each. Many are arranged in order to give strings of such structures each specialised to its task. Some help build the ear and are – as the fossils predicted – related to those responsible for gill slits, jaws and fish sensors. The vertebral column, too, is the product of such genes.

The simplest extant member of the greater vertebrate clan (fish, fowl and people included) is a small marine creature called the lancelet that spends most of its time buried in sand in shallow seas, where it filters food through its jawless mouth. It has a segmented body and, instead of a proper backbone, a simple stiff rod along its back. The animal's homeoboxes are arranged in the same order as its body parts. Many of its relatives (ourselves included) have four times or more as many copies of such structures. Their multiplication, followed by the divergence of the various copies, promoted the wild diversity of animals with backbones. The bony fish, which have doubled the number again, are the most variable of all in size, shape and way of life.

The vast variety of the crustaceans and their relatives – from spider crabs to fungus-like parasites to wasps – also emerges from their group's flexibility in development and they too have homeobox genes quite similar to our own. The variation upon a common theme reaches a peak among the cirripedes. Compared with their close relatives the lobsters they seem simple, for they have no obvious abdomen, and no more than a few jointed legs. Like snakes, lizards and whales the barnacles have lost limbs and, like birds compared with dinosaurs, have abandoned their back ends. The parasitic forms are even simpler. All this can be tracked to changes in their homeobox genes. The ancestral barnacle had ten, each of which has an analogue in geese and humans. The number of legs varies from species to species – and that variation is matched by the activity of two of the famous genes. The absence of an abdomen is due to a deletion of a group of homeobox genes similar to those that code for our own posteriors.

Darwin's 'unity of type' hence stretches from cirripedes to men and to the intimate details of the DNA itself. Homeobox genes draw together animals that at first sight show almost no resemblance to each other. Sir Robert Moray was, in a way, almost right about the barnacle's relationship to its eponymous goose: for perhaps those who saw a similarity between the two noticed their shared pattern of repetition; of vertebrae in the goose and body segments in the goose barnacle. If so, they were ahead of their time, for what might appear to be an accidental resemblance is proof of an ancient unity of form. Bird and barnacle each show how multiplication and divergence rule the world of life. They put paid to the absurd idea that complexity demands design or that evolution cannot generate information. The anatomy of those two sea-loving creatures, the pressure sensors of fish, the ear of an opera fan and large parts of the human genome are each a messy and expedient solution to a set of immediate problems. As Darwin noticed on the coast of Chile and as modern genetics can affirm, inelegant, redundant and wasteful as biology might be, it works well, but only as well as it must.

CHAPTER VIII

WHERE THE BEE SNIFFS

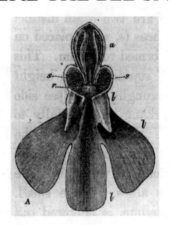

A gift of orchids is a statement of a gentleman's intentions towards a potential partner. A man willing to spend so much on his mate must be devoted indeed – or rich enough not to care, which comes to more or less the same thing. An orchid, with its extravagant flowers and a price tag to match, is a real test of his readiness to invest in a relationship.

The plants feel the same. Their Latin name, Orchidaceae, means 'testicle' after the unexpected shape of their roots. Orchids advertise their prowess with expensive and often bizarre blooms. So impressive are their carnal powers that the English herbalist Nicholas Culpeper called for caution when they were used as aphrodisiacs. In *The Descent of Man and Selection in Relation to Sex* Charles Darwin had shown how, in the animal kingdom, the battle to find a mate was as formidable an agent of selection as was the struggle to stay alive. Males, in general, have the potential to have far more offspring than do females – if, that is, they can fight off their rivals and persuade enough members of the opposite sex to play along with their carnal desires. Losers in the conflict reach the end of their evolutionary road for their genes go nowhere. Evolution as played out in the universe of sex is as pitiless as is that in the battle for survival. Sexual selection, as Darwin called it, can lead to rapid change: to the evolution of gigantic antlers, a vivid posterior or – for species interested in such things – gold watches and flashy clothes.

Later in his career, Charles Darwin examined the sexual struggles within the second great realm of life, the plants. He showed how the search for a partner can be as much of a challenge for them as it is for stags or peacocks. Plant reproductive habits were obscure and their mere existence often denied until the seventeenth century, but within a hundred years or so the basic machinery had been worked out. Flowers were both the home of the reproductive organs and an eloquent statement of erotic need. Darwin found that they evolved in rather the same way as an animal's sexual displays and were subject to the same forces of selection, which often achieved ends equally – or more – bizarre than those found in animals. In addition he discovered (although he found it hard to believe) that for orchids sex was full of dishonesty and discord, with all those involved ready to cheat whenever necessary.

Any botanical marriage is – by definition – more crowded than its animal equivalent, for a third party is needed to consummate it by moving male sex cells to the female. For some species, wind or water step in to help, but most flowers need a flying penis – a pollinator – to carry their DNA to the next individual (Ruskin, with his passion for the beauties of nature, strongly advised his young female readers not to enquire 'how far flowers invite, or require, flies to interfere in their family affairs'). Darwin himself saw how antagonism between the plant and animal partners is as powerful an agent of selection as is the process of female choice and male competition that gives rise to the peacock's tail. Flower and pollinator each become trapped into the embrace of the other and enter an evolutionary race that may end with the emergence of structures as unexpected, and tactics as devious, as anything in the animal world.

The interests of those who manufacture the crucial DNA and those who deliver it are quite different. From a female flower's point of view, or that of the female part of a hermaphrodite plant, one or a few visits by a winged phallus is enough to do the job (although the more callers she gets, the more choice she has of which sex cell to use). To beat its rivals, however, a male is forced to attract the distribution service again and again – and that can be expensive.

In his 1862 volume *On the Various Contrivances by which British and Foreign Orchids are Fertilised by Insects, and on the Good Effects of Intercrossing*, Darwin studied the divergence of interests between the two parties. He used the showiest and most diverse of all flowers as an exemplar. He found that 'the contrivances by which Orchids are fertilised, are as varied and almost as perfect as any of the most beautiful adaptations in the animal kingdom'. As well as an exhaustive account of the structure of the orchids themselves ('I fear, however, that the necessary details will be too minute and complex for any one who has not a strong taste for Natural History'), his work introduced the idea – much developed nine years later, in *The Descent of Man, and Selection in Relation to Sex* – that large parts of evolution depend on an ancient and endless sexual conflict that crafts the future of all those who are drawn in.

The war between flowers and insects became an overture to a wider world of biological discord. It has led to spectacular bonds between improbable partners, As it does, it reveals many of the details of the mechanism of natural selection, including its uncanny ability to subvert the tactics of any opponent. The orchids and their pollinators were, for Charles Darwin, an introduction to the dishonesty that pervades the world of life.

Pollination has attracted attention since ancient times. Both Aristotle and Virgil were interested in bees, but only because they made honey (or collected the stuff, for the Greeks imagined that it fell from the air: 'air-born honey, gift of heaven') rather than because they were essential for reproduction. The Egyptians, in contrast, understood that dates would not grow on cultivated palms unless male flowers were shaken on to the females. They used their slaves as pollinators. A diversity of other creatures has been called in to act as marital aids and quite often that duty drives their own evolution. Two hundred thousand insects (the male malaria mosquito included) are known to transfer pollen and their own vast radiation into a variety of forms began soon after the origin of flowers. From the tropics to the sub-Arctic, hundreds of species of bird are busy shifting genes. Some, such as humming birds, can almost never afford to stop as they need a constant supply of nectar to keep their tiny bodies at a high level of activity. Mammals are also involved, and a certain Ecuadorian bat has a tongue half as long again as its own body – in relative terms the longest of all mammalian tongues. It is coiled up in a special cavity in its chest, except when the animal is feeding. An African tree is even adapted for pollination by the giraffes that browse upon its leaves. In Australia, too, marsupials have taken up the job for the honey possum has lost many of its teeth and gained a long tongue. The sugar glider – a marsupial that floats through the air from flower to flower – is much the same.

Plants want their go-betweens to be cheap, trusty and eager, while pollinators would prefer to be fat, wanton and as idle as possible. The flower shows that a reward is on offer while the other party must decide whether the hard work needed to get it is worthwhile. The struggle between the two parties leads to the evolution of displays that dwarf the efforts of any animal. A bunch of flowers is an advertisement – a silent scream from the sexually frustrated. Like all advertisements it attempts to reassure those who see it that a high-quality product is on view. In commerce, as in life, the temptation to cheat is never far away; to make false promises with no reward, or to take the prize and fail to complete the task.

Plants and animals make signals of many kinds. They advertise their qualities as a mate, their willingness to fight for territory or food, or their ability to escape from a predator who might as a result be dissuaded from bothering to attack. One surprise is that the signals are so often honest when the reward for dishonesty is so high, be it in the form of sex, food or safety. Some signs are direct and impossible to fake: large tigers make scratch marks higher up a tree trunk than can their smaller rivals and can as a result hold bigger territories. Often, though, the information is indirect. Thus, a black and yellow wasp warns predators about its dangers without the need to sting all of them.

Such secondary signals, too, are sometimes pricey and hard to simulate. Giant antlers, vivid tails or spectacular blooms can be made only by those who can afford them: the healthiest, the sexiest or the super-aggressive. Most of what we interpret as the joys of nature costs a lot, for a stag may die in battle, and a male nightingale loses a tenth of its body weight after a night spent singing in the hope of sex. Testosterone itself, that signifier of masculine identity, is costly in many ways. It suppresses the immune system, so that a red deer male in sexual frenzy is open to attack by parasites – and if he can keep roaring in spite of his tapeworms he might have particularly fine genes. Elephants go even further. Now and again, one falls into a state of 'musth', in which its testosterone level goes up by fifty times. The agitated beast becomes very aggressive, and a small animal will fight even to the death against a larger rival.

Flowers, too, are not cheap. Orchid fanciers pay tens of thousands of dollars for prize specimens and the trade as a whole has a worldwide turnover of several billion. The orchids themselves invest far more of their limited capital into sexual display than does the most avid gardener, for if they do not their genetical future is over. The cost of sex to each orchid and to those that market them is manifest in the fact that, in the world of the garden centre, many of the specimens are grown from cells in culture rather than by persuading the plants to go through the expensive ritual of sex.

Orchids and other flowers are, like the peacock's tail, animated billboards that advertise sexual prowess. For all signals, two parties are involved: those who transmit a message and those who receive it. A system of checks and balances tests whether the information is accurate; that those with the biggest antlers or brightest blooms really are the fiercest or most generous. The system is always under test by potential fraudsters at both ends and sometimes cheats get in. Often, they do well. For insects, black and yellow is no more difficult to manufacture than is brown or blue – and a whole group of harmless flies does just that, with bright stripes that make a false claim of a waspish nature. That costs the wasps a lot when a hungry bird attacks under the assumption that the pattern advertises good taste rather than potential danger.

Such swindlers also flourish in the botanical world – and in orchids most of all. To his considerable surprise, Charles Darwin found that among those elegant flowers dishonesty pays. Many of his specimens had gorgeous displays, but gave no payment to their pollinators. He found it hard to believe that Nature could be so fraudulent or that insects were so foolish as to fall for 'so gigantic an imposture' and suggested, wrongly, that his plants had an as yet undiscovered reward. His finding throws light on a question that he posed but failed to solve: how can natural selection favour the dishonest? The orchids give part of the answer.

The battle for sex is a war of all against all. It may end in an arms race; a tactical struggle in which every move made by one party is countered by the other. Sometimes, as in the Cold War, each antagonist is forced into massive investment, and, as in those days, negotiation may end in stalemate. To an untutored eye that may look like peace, but it is in truth no more than battle deferred. The orchids, beautiful as they are and exquisite as their adaptations to the needs of their pollinators might be, show such a struggle hard at work and show how propaganda – false information – is useful in both love and war.

Many of Darwin's own observations were made on the 'Orchis Bank', close to his home, where he found eleven species of the plants. As he noted with a certain pride, 'no British county excels Kent in the number of its orchids', but he also studied specimens sent from all over the world. He soon saw how the conflict between plant and pollinator had led to change. He speaks of an orchid, 'the *Angræcum sesquipedale*, of which the large six-rayed flowers, like stars formed of snow-white wax, have excited the admiration of travellers in Madagascar'. It had 'a whip-like green nectary . . . eleven and a half inches long, with only the lower inch and a half filled with very sweet nectar. What can be the use, it may be asked, of a nectary of such disproportional length? . . . in Madagascar there must be moths with prosbosces capable of extension to a length of between ten and eleven inches! . . . As certain moths of Madagascar became larger through natural selection in relation to their general conditions of life . . . those individual plants of the Angræcum which had the longest nectaries . . . and which, consequently, compelled the moths to insert their prosbosces up to the very base, would be fertilised. These plants would yield most seed and the seedlings would generally inherit longer nectaries; and so it would be in successive generations of the plant and moth. Thus it would appear that there has been a race in gaining length between the nectary of the Angræcum and the proboscis.' In 1903, that long-tongued insect, product of an endless contest with its plant, was at last discovered and named as Morgan's Sphinx Moth. A long conflict of interests had forced both parties to adapt themselves to each other's demands.

As the sage of Down House collected orchids from the fields

and heaths around his comfortable home and examined the specimens sent to him from afar, he became more and more impressed
by the ingenuity of the ways in which as they pass on pollen:
'Hardly any fact has struck me so much as the endless diversities
of structure, – the prodigality of resources, – for gaining the very
same end, namely, the fertilisation of one flower by the pollen
from another plant.' He glimpsed but a small part of the game
played by all plants as they fulfil their sexual destiny.

As Darwin showed, nine years after the orchid book, in *The
Descent of Man, and Selection in Relation to Sex*, a male peacock's
flashy rear says nothing about the merits of tails, but a lot about his
status as a high-quality mate who can afford a gorgeous adorn-
ment. The same is true of plants. More food allows them to make
more blooms and to proclaim their excellence to a larger audience. To remove a few flowers may also allow them to grow more
fruits, as proof of how expensive it is to be attractive. The brightest and most generous individuals get more pollinators and pass on
more of their genes, which promotes yet more brightness and
generosity in the next generation and, almost as an incidental,
leads to an outburst of diversity as the balance of sexual advantage
species in different lineages.

Orchids belong to the great subdivision of the flowering plants
that generates just a single leaf as the seed germinates. It contains
the grasses (crops such as rice included), bananas, tulips and more.
The orchids themselves are among the largest families of all for
only the group that contains daisies and sunflowers possesses more
species. Around twenty-five thousand different kinds are known –
about an eighth of all plants with flowers – and no doubt many
more remain to be discovered. Britain has just forty-six native
kinds, several of which are rare.

Because orchids are so attractive they are important in the conservation movement (and cynics call them 'botanical pandas').
They may look fragile but many are tough. Their capital lies in
the wet and cool hill-forests of the tropics, and a third of all

known species are found in Papua New Guinea. Plenty more live in the Arctic or in temperate woodlands, fields and marshes. They grow on the ground or high in the branches of trees, or on rocky slopes and grasslands. A few live underground and never see the light of day. In some places the plants are short of water and, like cacti, develop thickened stems or tubers to store a reserve. Some have leaves as big as their relatives the bamboos while a few are parasites with almost no foliage at all. Others, such as the vanillas, make vines twenty metres long. Some kinds are tiny, with a flower head that would fit on the head of a pin, while the flower of a certain tree-dweller from New Guinea is fourteen metres around and weighs about a thousand kilograms (a specimen caused amazement at the Great Exhibition in 1851). Plenty of others have multiple displays several metres long. A few have opted out of the endless and expensive conflict and are pollinated by the wind while one Chinese kind has abandoned the whole business of sex and indulges in a strange internal dance in which its male element curves backwards and inserts itself into its own female orifice.

Some of the flowers are simple. They are dark and look rather like the entrance to a burrow, which attracts a bee to come in for a snooze and pollinate as it does so. Many others use far more elaborate tactics. Some are perfect six-pointed stars while others resemble a glass-blower's nightmare with fine tendrils that hang together in delicate and lurid bunches. Yet others look as if they are moulded from thick pink plastic. The flowers are scarlet, white, purple, orange, red or even blue. One species is pollinated by a wasp. It generates a chemical identical to that emitted by a leaf chewed by grubs − the wasp's favourite food. The wasp as it visits gets not a meal of tasty flesh, but a load of unwanted pollen. For those over-impressed by the beauties of botany, certain orchids smell like putrid fish to attract carrion-feeding flies.

The biological war between flower and insect, like the whole of evolution, involves an endless set of tactics, but no strategy. It has produced a vast variety of blooms, each of which evolved in a manner that depends on the preferences of their pollinators and on what turns up in the form of mutations. Darwin noted what strong evidence the orchids were against the then common notion that the beauties of nature emerged from some kind of plan: their structures 'transcend in an incomparable degree the contrivances and adaptations the most fertile imagination of the most imaginative man could suggest'. They were another weapon in the battle against the idea of design, a 'flank movement on the enemy'.

However remarkable the details, all their flowers are based on the same fundamental plan. It resembles that of the distantly related, but simpler, lily (and Goethe himself, with his interest in botany, described orchids as 'monstrous lilies'). The parts are arranged in threes, or multiples of that figure. The central lobe is often enlarged into a coloured lip which acts both as a flag to attract insects and as a landing strip that allows the visitors to reach the sweet reward at its base. Often, the flower rotates to turn upside down as it develops. The male organ sits at the end of a long column and the male cells, the pollen, are not powdery as in other plants but instead are held together in large masses, with up to two million minute grains in each. They are covered with a sticky secretion that can attach the whole lot to an insect. The female part lies deeper within, on the same column as the male. Once fertilised, the orchid may produce thousands of tiny seeds in every capsule – no more than a minute proportion of which have any hope of success.

When the pollinator enters, some means is found to attach male sex cells to it. Many orchids have a spring-loaded mechanism that fires a mass of pollen in the right direction. It sticks on with powerful glue. As Darwin found by stimulating the flowers with a pencil, the stalk of the transferred pollen sac quickly dries out and the mass of male cells takes on a more vertical position, just right to fertilise the female part of the next plant visited. In a few kinds, should the mass miss its target, its energy is enough to shoot it for a metre away from the plant (the pollen is 'shot like an arrow which is not barbed'). The blow is unpleasant enough to cause an insect that has been hit to concentrate, if it can, on the female part of the flowers it visits subsequently, which is a real help to the male who scared it off. In other species, the pollen masses crack open to the sound of a buzz like that of a particular species of bee. Yet other kinds have a see-saw that tips the insect on to the crucial male cells. It pays the plant to do the job as well as it can, for many orchids are limited in their ability to reproduce by a short-age of visitors.

Once the pollinator has been enticed to arrive it expects to be repaid. The first reward of all, in the earliest flowers, was pollen itself, which is expensive as it contains lots of protein. Even so, plenty of orchids still provide a solid meal made up of the stuff, or of bits of tissue that resemble it. Others give nectar, which is simpler and can be provided in very dilute form. Honey bees, for example, must extract the sugary liquid from several million flowers, a few of which may be orchids, to make a kilo of honey.

Some botanical entanglements are intimate indeed. Certain bees are so tied to their partners that their own sex lives have come to depend upon them. They obtain their sexual scents – their pheromones – from an orchid flower and without a visit they are unable to mate. The pheromones may have more than a dozen ingredients. As in the Chanel factory, the bees practise 'enfleurage': they mix an odoriferous base taken from the plant with an oily substance of their own that helps the smell to persist. A special grease is smeared on to the flower and the sexual mix transmitted to pockets in the bees' hind legs. The habit evolved from the insects' ancient habit of marking their sexual readiness – like dogs around lampposts – with scents extracted from flowers, rotten wood and even from faeces.

The battle is not evenly matched, for the insects themselves – many of whom visit a variety of plants – are under less pressure to retain an accurate fit with their partner than are the flowers. The orchids evolved well after insect pollination began and have had to adapt to the needs of their partners, rather than the other way around. Some insects – many bees included – are quite catholic in their tastes and some orchids are indifferent as to who moves their pollen, as long as somebody does. A few species are visited by more than a hundred different insects, while only around half of all orchids are more or less faithful to a single pollinator. To become too closely connected to a particular insect is risky. Darwin himself speculated that the giant orchid of Madagascar would disappear if its specialised pollinator died out, and he may have been right.

The orchids face a higher risk of failure if they cannot find a pollinator than do animals in the same predicament, for an insect can always try another kind of flower if its prime source of food becomes too rare or too mean. Many flowers – those of orchids included – are in fact visited by several pollinators, even if particular species do tend to concentrate on similar insects; on long-tongued bee-flies and long-tongued flies, or on tiny bees, flies and beetles, each of which picks up the pollen on its legs. Even the bees that pick up their own sexual scents from an orchid are less dependent than they seem. A certain South American species has become naturalised in Florida, where its host does not grow. It finds its chemicals instead in aromatic plants such as basil and all-spice when it chews their leaves and extracts the smelly substances. The bee pollinates a wide variety of local plants, which reciprocate with nectar rather than with an aphrodisiac. Once again, the insect has more freedom of action than does its partner.

As a result, the two parties are often less entangled than Darwin imagined. A shift in one is not always matched by an equivalent move by the other, with deeper flower trumped by longer tongue. Molecular trees of plants and pollinators suggest that the insects have instead often switched to species with shallower flowers from which nectar can be sucked with less effort.

The orchid's ability to force its ally to serve its selfish interests is further limited because such gorgeous beings are often rare and scattered among other species. Make life too hard and the insect will sip elsewhere. Infidelity by the pollinator is bad news for the orchid as it may fail to export its own genes and in addition it may get pollen from the wrong species. Nevertheless, not all the pollinators have been promiscuous, for fossil water lilies from ninety million years ago have flowers quite like those of their modern ancestors as evidence that their association with beetles is ancient indeed.

The pressure for sex often causes natural selection to run away with itself. Like many showy animals, birds and butterflies included, there are lots of different orchids. Twenty-five thousand kinds are known, compared with no more than a hundred or so species of wild roses (which are happy to attract almost any insect that might pass by). Most of the barriers to gene exchange among the orchids are held in the brains of their pollinators. As a result, the fertile minds of gardeners have been able to generate thousands of hybrid forms by getting round the ancient bond between flower and insect with a simple paintbrush. Their success shows how fine the balance of barriers to the movement of DNA must be. A tiny shift can change the equation of flower and pollinator and make a new species. In some cases a mutation that changes colour from a hue attractive to bees to another favoured by birds has started a new species in a single step. In the same way, in orchids pollinated by scent-seeking bees, a subtle shift in the proportions of each constituent can attract different kinds of bee, which means that physically identical plants may in fact be distinct entities that never exchange genes.

Orchids bolster Darwin's case that species arise through the action of natural selection and he soon realised that their diversity had been driven by the vagaries of insect behaviour. He was much less certain of the origin of flowers themselves, which he called 'an abominable mystery' and a 'perplexing phenomenon'. The mystery has been cleared up and the orchids have helped.

Plants colonised the land more than four hundred million years before the present. Those pioneers had no flowers and neither did the huge forests of giant ferns that covered large parts of the planet a hundred million years later. The fern forests declined and the dinosaurs flourished for an age in a flowerless world. Not until the first flowers of all, perhaps a hundred and fifty million years ago, did the conflict between insect and plant begin. It led to an explosion of change in both parties. Their joint transformation was spectacular, for more than three hundred thousand species of flowering plants have evolved, together with several times that number of insects.

The oldest fossil flower comes from a famous bed close to the estuary of the Yellow River in China. It dates from around a hundred and twenty-five million years ago, at the time when the white cliffs of Dover were being formed in a shallow sea. It looked rather like a water lily and floated in fresh water with its small flowers above the surface. For tens of millions of years such structures remained modest, but sixty-five million years ago – just as the dinosaurs left the stage – the world burst into bloom.

The orchids played their part in beautifying an unpeopled world. A distinctive pollen sac attached to a stingless bee has been found in twenty-million-year-old amber from the Dominican Republic. That orchid's modern relatives use just the same group of insects to transfer their male cells. The molecular clock suggests that orchids as a whole originated around the time of the extinction of the dinosaurs. Their massive radiation happened just after that memorable event and was accompanied by parallel change in the insects that pollinate them.

The great blooming was evidence of an early skirmish in the war between orchid and insect. Conflict between plants and pollinators has gone on ever since. It is expensive and never more so than when it escalates. The Soviet Union collapsed under the financial pressures of its attempts to match the power of the Americans and for centuries Britain and France spent a third of their wealth in mutual conflict. In war, as in love and business, lavish display is a test of merit. A military parade intimidates the enemy and a costly publicity campaign is a sign of a high-class company. The medium becomes the message, the powerful stay in charge, cheats go bankrupt and, for most of the time, truthful ostentation prevails. The best signals are too expensive to copy, which is why McDonald's sues anyone who imitates their golden arches and why Japanese Yakuza gangsters cut off their fingers.

The interaction between plants and pollinators is a matter of economics – and economists have been quick to notice. Signalling theory tries to explain how decisions are made when the information available is less than perfect – what used car to buy, who to hire for a job, what flower to visit. One test is to look for a reliable sign of quality, whatever it might be. An applicant for a job in a bank might have a first-class degree in genetics. Useless as that certificate might be to a prospective financier, it is at least an honest (and expensive) statement of overall merit. The system works well – as long as everyone is honourable. Sometimes they are not. Straight fraud – a forged Harvard degree – can often be picked up but what of a parchment from one of the many bogus universities that nowadays advertise their wares? The University of Dublin sounds respectable but is a website. How can employers tell Redding University (an American degree mill) from the University of Reading (a respectable institution to the west of London)? Thousands of people now have such qualifications and if too many degrees turn out to be false then the whole machine breaks down. The risk is real. Nine-tenths of the 'Tiffany' jewellery on sale on eBay is fake, and Tiffany & Co. has spent millions in attempts to shut down the sellers, who cause huge damage to its brand. If the bogus continue to prosper at the expense of the genuine, the entire jewellery market may collapse.

A study of the economic implications of such false signals (or 'asymmetric information', as financial experts call it) won the Nobel Prize for Economics in 2001. Plants and animals have done such sums for years. Most of the time, they get it right and honesty more or less prevails. Sometimes, the cheats get in, for if the reward is large enough and the penalty for swindling not too stringent, natural selection can favour sharp practice. The temptation to invest in display rather than product means that the price of sex is eternal vigilance. Some orchids – like some traders – allow others to pay for the publicity while they double-cross their pollinators. Life at the top faces a constant challenge from fraudsters.

Plenty of pollinators, too, are duplicitous. Insects gnaw into a flower to gain a reward at minimal cost while humming birds can poke a hole in its side to do the same. Even legitimate pollinators like honey bees become robbers at once when someone else has broken in. For them, dishonesty pays and they turn to it whenever they get a chance.

The flowers have hit back. What they offer may be quite different from what they promise. Orchids have a wide range of lures. Some subvert their pollinators' desires with blossoms that resemble female insects. The flowers are larger than real females, and may emit a hundred times more of their attractive sexual scent. The male bees or spiders – understandably – try to copulate with their spurious brides and in their failed attempt to pass on their own DNA do the same job for the plant. Their amatory experience is futile but intense, as many of the befuddled males produce copious amounts of sperm that costs them a lot to make and goes nowhere.

Darwin found it hard to believe that a bee could be so stupid as to frot a flower but, in the world of sex, stupidity can pay. A naive male bee faced with females in short supply, as they often are because males emerge earlier in the season than their partners, is well advised to travel hopefully because he might arrive; he should copulate with anything that looks even a little like a member of the opposite sex on the off-chance that, now and again, he will be lucky. The bees oblige and, most of the time, the orchids win. Other orchids exploit the aggressive, rather than the amatory, instincts of their pollinators. They mimic a male insect rather than a female – which annoys the local territory-holder who tries to drive out the supposed intruder and pollinates as he does so.

Other kinds exploit the greed, rather than the lust, of their visitors. They advertise not sex but a free meal but again, they provide nothing. That baffled Charles Darwin. It was 'utterly incredible' that 'bees . . . should persevere in visiting flower after flower . . . in the hope of obtaining nectar which is never present'. He suggested instead that the empty flowers had hidden reserves, which the insects would reach if they made a hole and sucked the plant's juices.

Life is less honest than he imagined and the flowers were in fact cheats. About a third of all orchids act in this underhand way – flashy signal, but no food reward. Some other plants do the same, often with a few 'cheater flowers' on an individual in which most are honest, but the orchids are the real confidence tricksters, for they make up nine-tenths of all flowers known to fool their visitors. DNA shows that the habit has arisen again and again within the group – but it does not always pay, for some orchids that now provide a generous recompense to their visitors have evolved from species that once led a dishonest life.

Often, such false flowers are – like the harmless flies that look like wasps – mimics, with a resemblance, more or less accurate, to other local plants that do make a reward. They flaunt a badge of quality such as bright colour to attract an assistant on the cheap. Some work hard to fool their visitors and are uncannily similar to a particular model in shape and colour. Certain Australian kinds, for example, look like mushrooms and are pollinated by fungus gnats in search of a place to lay eggs. A few even make small orange and black spots on their flowers which attract aphid-feeding flies that see the spots as potential prey. More often, their displays are no more than general statements of reward that attract a variety of insects. The parasite joins a whole guild of locals in which the various species share a resemblance and attract about the same mix of insects. Honest plants pay the price when insects avoid them after an anticlimactic experience with a cheat. Some orchids are doubly duplicitous for individuals vary in colour, one from another, which allows them to parasitise a wider range of victims.

The cheats tend to grow scattered among their hosts, for a group of fraudsters close together is soon detected by the pollinators, who move away to a more worthwhile patch. They do best at fooling insects that have just emerged into the wicked world and have not yet learned to detect double-crossers. As a result such orchids tend to flower in the spring rather than later in the year. but in many cases a shortage of pollinators foolish enough to revisit a dishonest plant force it to make a long-lasting flower and pollen that, unlike that of most of its fellows, survives for weeks or months. A certain Australian orchid uses the opposite strategy, for all the plants open on the same day of the year, giving the pollinators no time to learn about the gigantic fraud being perpetrated upon them.

Experienced insects soon become cynical for they move away faster – and fly further – from empty flowers than from those with nectar. The dishonest orchids may reap a subtle benefit from their disappointed visitors, for the still hungry insect may buzz off to a new individual, rather than shifting its attentions to a second flower on the same plant. Such behaviour cuts down the chance of self-fertilisation.

Orchids may be the real experts, but plenty of other associations between plants and pollinators have been subverted by natural selection. Wild peas and beans often make nutrient-rich rewards that attract birds to spread their seeds but some of their offerings contain nothing of value although they look like a tasty meal. Yuccas – those spectacular flower spikes of the American deserts – are pollinated by a certain moth, who carries a bundle of pollen to the female, inserts it in the right place and then lays her eggs within the flower. When they hatch, the larvae feed on the seeds and once adult fly off to pollinate another yucca. Close relatives of such moths, though, eat the seeds without bringing pollen.

The fraudulent orchids and their fellows among the pollinators were an introduction to a wider world of sexual dishonesty that has emerged since Darwin's day. When it comes to the need to pass on DNA on the cheap, animals are just as devious as are plants (although not many can match the orchids, in which an entire species may transmit its genes by Machiavellian means). Plenty of animals are bullies who boast of powers that they do not possess, or swaggarts who claim sexual prowess but in truth are feeble. An ability to roar even when filled with parasites or a readiness to die in the battle for a mate is hard to fake but, as in the orchids, a dependable statement of quality can sometimes be subverted.

Many male insects use a gift of food to persuade a female to copulate with them. Dance flies, hairy-legged predatory insects of wet places, make swarms in the mating season. In some species, each male brings a gift of a dead insect larva, and mates with his female while her attention is diverted by the meal. Once the bribe has been eaten, the male is pushed off. Other species prolong the sexual experience, for they wrap the gift in a silk purse, which the female must open before she can eat. At once, a chance to cheat presents itself – and it has been seized. Some male flies make elegant and complex purses that take a long time to open, but – like a dishonest orchid – are empty, or contain a desiccated corpse. By the time the female finds out, she has been inseminated. Fireflies are just as devious. Males bring a gift, a sticky mass of nutritious gel that goes with the sperm and is soaked up by their mates. Those who can afford more of the stuff make a longer flash and attract more females. A successful male soon runs out of energy. Some cheat, with a long flash and no reward – but they take a risk, for a certain predatory firefly uses the burst of light to find its prey. A false flasher risks death every time he exposes himself.

Darwin's perplexity about the dishonesty of orchids opened the door to a whole universe of evolutionary discord. Many creatures are happy to lie in the race to pass on genes. The conflict extends beyond plants and pollinators, to predators and prey, pathogen and host or men and their domestic animals, all of which are locked into an endless – and often joyless – conflict. Such ancient disputes explain why the Irish had a potato famine, why some diseases are virulent and others not and why the Argentinian Lake Duck has a corkscrew-shaped penis longer than its own body.

Sexual dishonesty is widespread. Birds are at it all the time. Many species appear to live as faithful pairs, but paternity tests show that the majority are happy to cheat and that half – or even more – of the eggs of a particular female are the scions of another male, often an individual more dominant than their regular partner. Mammals are even more devious. The joys of paternity-testing reveal that a male mammal's sexual displays are often subverted: a feeble individual can sneak in when the top stag is preoccupied with display and insert his own genes with no need for a huge investment.

Monogamy is rare, for not more than one mammal species in about twenty (some humans included) appears to indulge in it. Even some classic examples of reproductive honesty are in fact cheats. The prairie vole seems to stick to his mate through thick and thin and helps raise the young. Their happy marriage is based on a certain hormone. On his wedding night a surge of the stuff kicks in and appears to tie the male to his partner for life. A director of the US government's family planning program saw the vole as proof that sex before marriage disrupts brain chemistry and leads to divorce. The hormone, he says, is 'God's superglue'. It bonds partners together and, said the politician, it does the same for society (and also proves that abstinence is the finest form of contraception). The gene that picks up the hormone in the bloodstream comes in several forms in humans, too, and – in Sweden at least – men who bear two copies of a certain variant are less likely to be married or, if they are, have a more difficult relationship than do others.

The cold eye of the paternity-tester has now fallen upon the private life of the prairies. DNA cannot tell a lie – and it shows that beneath the vole's upright social habits lies a dark sexual universe. One in five of the young of each pair is fathered by a male other than the marital partner and around a quarter of all males and females have sex outside the household. Voles are socially faithful, but sexually fickle; happy to cheat, but quick to forgive. Foxes are even more dishonest, for more than three-quarters of their cubs are fathered by a stranger.

Darwin was surprised by the reproductive fraud he found among orchids – but refused to accept that the same could be true for mammals, for humans least of all. In his view of sexual selection, males might be promiscuous or even crafty, but females were monogamous; they chose, and males competed for their attentions. Part of that Puritan philosophy was due, perhaps, to the social climate of the time and his reluctance to shock the female members of his household. In modern society, in contrast, the concept of dishonesty in sexual relations has almost disappeared as most liaisons consist of longer or shorter periods of serial monogamy, accepted by both parties. That shift shows the flexibility of human behaviour and how hard it can be to draw any worthwhile lessons about our own private lives from those of other mammals, let alone of flowers.

Even so, there has been plenty of reproductive dishonesty in our own history. Casanova, himself of uncertain paternity, posed as a soldier, a doctor, a diplomat, a nobleman and a sorcerer to gain the favours of an admitted hundred and twenty women (plus, more than likely, many more). He was a great lover, and a better liar, even if, according to a contemporary, he 'would be a good-looking man if he were not ugly'. His wit, rather than his looks, charmed his way into the bedroom.

Now, the chance for deceit has been much improved by technology. No longer does a hopeful male need to display his talents directly; instead he can say what he chooses about his looks, his education and his wealth on an online-dating site. There he has no fear of detection, at least until his first appointment with a prospective mate. Tens of millions of people use such sexual aids, and millions of liaisons (many ending in marriage) have emerged from a digital romance. Even so, nine out of every ten users – and women more than men – are convinced that the world of electronic eroticism is filled with cheats, with dirty and decrepit Casanovas who present themselves as young lovers in the hope of reproductive success on the cheap.

In fact, such suspicions are misplaced. Surveys of on-line daters show impressive levels of accuracy in their descriptions of themselves, for almost all say something close to the truth about age, body build, wealth, education, politics, marital history and more (admittedly, men tell slightly more lies about their income and women about their weight). The daters disapprove strongly of anyone who did not live up to their claims on a first meeting and swore that they would go no further with them. Deception is not an effective sexual strategy. For men and women, honesty pays and the fraudulent are rejected as partners as soon as they are detected.

In the dating game, on the other hand, there are few disappointments that a bunch of orchids will not put right.

CHAPTER IX

THE WORMS CRAWL IN

The fields of Britain are criss-crossed by earnest men with metal-detectors. Despised by archaeologists for the damage they cause, the 'discoverists', as they call themselves, have found thousands of coins, swords, belt buckles and the like. Some of the objects were hidden, or buried by their owners in times of danger, but most simply sank from sight. Why?

Charles Darwin, as usual, got it right. The past had been entombed by worms. He hymns their praises in his last book, *The Formation of Vegetable Mould, through the Action of Worms, with Observations on their Habits*: 'The plough is one of the most ancient and most valuable of man's inventions; but long before he existed the land was in fact regularly ploughed, and still continues to be thus ploughed by earth-worms. It may be doubted whether there are many other animals which have played so important a part in the history of the world.' His literary swansong discusses the anatomy and habits of such creatures, their intellectual life (such as it is) and, most of all, their ability to disturb the surface of the Earth, to aerate, turn over and improve the soil, and to sink any object that lies upon it. At the time its author claimed that he had produced no more than 'a curious little book' on a matter that 'may appear an insignificant one', but the ravages of the plough since it was invented thousands of years ago and the damage done to the surface of our planet by today's agriculture mean that the work of the worms is crucial not just to the history of the world but to its future.

The power of such small beings to seal the fate of objects far larger than themselves shows, once again, the huge consequences that can emerge from what may appear to be the trivial efforts of Nature. Darwin was aware of the potential of the worm as proof of the might of slow change; as he said of their efforts: 'the maxim *de minimis non curat lex* does not apply to science'. They were the final test of his obsession with the cumulative potential of the small and he was proud of his results. He dismissed the arguments of a Mr Fish, who denied the animals' talents, as 'an instance of that inability to sum up the effects of a continually recurrent cause, which has often retarded the progress of science, as formerly in the case of geology, and more recently in that of the principle of evolution'.

The elderly savant's attraction to such creatures had started long before he thought of science. In his autobiography he notes that, as a child, he had been so upset by their contortions when impaled on fish-hooks that, as soon as he heard that it was possible to euthanise them with salt and water, he never again 'spitted a living worm, though at the expense, probably, of some loss of success!' His later studies introduced a new world beneath our feet, gave life to the idea of animals as a geological force and, as an incidental, showed how even simple animals have a rich mental life of their own. His work became the foundation of a science which has now, almost too late, noticed the dire state of the world's vegetable mould and has begun to do something about it.

In 1837, just a year after the *Beagle* voyage, Charles Darwin presented a paper on worms to the Royal Geological Society. Later he published a few notes on the subject, which occupied him at odd moments for forty years. At last, at the age of seventy-two, he wrote *Vegetable Mould*, which was published at nine shillings in 1881, just six months before his death. The book was received with what he called 'almost laughable enthusiasm' and sold nearly as many copies in its first few years as had *The Origin*.

Soil is where geology and biology overlap. Adam's name comes from *adama* – the Hebrew for soil – and Eve from *hava*, or living: an ancient statement of the tie between our own existence and that of the ground we stand on ('*Homo*' and 'humus' also share a root). The epidermis of the Earth is no more than around one part in twenty million of its diameter while our own skin, in contrast, is about a five-thousandth as thick as the average human body. Leonardo da Vinci wrote that 'We know more about the movements of the celestial bodies than of the soil underfoot', and until *Vegetable Mould* that was still almost true.

Since then, earthworms and their relatives have been stud-ied by geologists, ecologists, molecular biologists and many others. Archaeologists, too, have reason to be grateful for their efforts, for without the animals our insight into history would be far less complete than it is, for most of the evidence left by our ancestors would not be buried but washed away. More important, perhaps, without worms we would starve.

Vegetable Mould built upon an observation Darwin had made as a young man. Twelve months after his return to his native island from his famous voyage, he visited his uncle – and future father-in-law – Josiah Wedgwood, at Maer Hall in Staffordshire. Wedgwood took him to a field upon which had been scattered, fifteen years earlier, a mass of lime, cinders and burnt marble, the detritus of his Etruria pottery works nearby. The material had, over that period, been covered by a layer of earth. Wedgwood suggested to his nephew that perhaps worms had done the job. The young scientist agreed, but saw this at first as little more than a 'trivial gardening matter'. In time, as the notion that – for both rocks and flesh – small means could give rise to large ends grew in his mind, he saw in those humble creatures a real chance to experiment on the measured actions of Nature.

Darwin continued to study the animals as he travelled across England with his wife and children. They were not the only tourists of those days. The Victorians were fond of excursions, and many were, like the modern discoverists, anxious to cart off relics for their own delight. In 1877, in a brief respite from ill health, Charles took his wife to visit Stonehenge. He dug pits around several of the 'Druidical stones', as the monoliths were then called, and noted that even the largest had been sunk several centimetres deep by the worms. Emma worried that her husband might have sunstroke as he sported with the relics, and she recorded her conversation with the site's guardian, 'an agreeable old soldier'. 'Sometimes,' the venerable trooper told her, 'visitors came who were troublesome, and once a man came with a sledge-hammer who was very troublesome to manage.'

A few years earlier a hammer and chisel had been provided at Stonehenge for the use of those who wanted a curio of ancient times. Their intellectual descendants still agitate the nation's soil. A hobby that began with chunks chipped off monuments has become an electronically powered craze, with, at its peak, almost two hundred thousand enthusiasts in Britain (they include Bill Wyman of the Rolling Stones, who markets his own metal detectors for 'Treasure Island UK'). For a time the 1960s Prime Minister, Harold Wilson, was an honorary patron of the detectorists' organisation. The numbers are well down from those frenzied days and in most of Europe the practice remains illegal, but in 1997 the British government bowed to reality and changed the law to reward those who report their finds; they are, said the minister for culture of the time, 'the unsung heroes of the UK's heritage'. The Portable Antiquities Scheme, as it is called, applies to England and Wales alone, for Scots must still give up their treasures to the Crown. South of the Border, the number of objects reported has risen from fewer than a hundred per year to thousands. The scheme now lists more than three hundred thousand items. Recent triumphs include the discovery of a four-thousand-year-old gold cup at Ringlemere, in Darwin's home county. Its finder shared a £250,000 reward.

The Ringlemere cup was concealed by men anxious to placate the gods but many other objects have been hidden by humbler creatures with simpler motives. Electronic sweeps of the fields around Down House reveal many coins, necklaces, buckles and the like. Vast numbers more remain, no doubt, to be uncovered. They were buried by worms as they searched for shelter and for food.

The earthworm has undoubted charm. It belongs to a group known as the annelids, which include the leeches and lugworms and is related to less agreeable creatures such as the parasites that cause elephantiasis in tropical Africa. The creatures are more distant kin of snails and slugs. Their ancient roots are best revealed by patterns of DNA similarity, as soft-bodied creatures do not leave many fossils (even if the remains and the tracks of a few primitive annelids are found as far back as the Cambrian). Today, three thousand or so species are known, and, given our ignorance of tropical nature, many more must remain to be found. Most are small and unassertive, but a certain Australian kind grows to three metres long and ejects a jet of fluid half a metre into the air when annoyed. Britain has a just over a couple of dozen sorts while France has six times as many. A rain forest has far more.

A 2005 survey at Down House, in the woods by the famous Sandwalk – the site of its owner's regular stroll – and in a nearby meadow in which Darwin had noted that stones were soon buried by worms, revealed that his home was still a hotbed for the creatures. Nineteen of the twenty-eight British species were found there or nearby. The most abundant species nowadays, and no doubt in Victorian times, was the black-headed worm, which is smaller than the familiar lobworm found in city gardens and used as bait by fishermen. The animals were most abundant in the kitchen garden, probably because of its many decades of fertiliser and a strict ban on pesticides.

A worm is an animated intestine. The body is divided into segments, each with an outer layer of muscle that encircles it and an inner muscle sheet that runs parallel to the axis. Each segment bears a simple kidney with a series of even simpler hearts distributed along the animal's length. The body is hollow and filled with fluid and down the centre runs a long digestive tube. Many species have internal glands filled with lime – calcium carbonate. Some have coloured blood while others are almost transparent. Certain kinds smell of garlic, perhaps to put off predators.

The skin is covered with stiff spines that help the creature move through the soil as it eats its way onwards, pumping out waste from the rear end as it goes. In some kinds, the slime made as it burrows hardens into a solid wall that keeps a track open for a possible return while in others the soil collapses behind the questing worm as it travels. Some worms live on or just below the surface and in leaf litter, while others hide deeper, sometimes six metres down. A few prefer rotten wood. Those most important to farmers roam the top metre or so of soil. Some species reuse their burrows while others set out instead to build new homes. The common lobworm makes a single excavation, with one or two branches, while others make a network with several exits.

Most earthworms spend most of their time at rest in their underground fortresses and venture forth only when conditions are suitable. In winter they dig down and hibernate and in dry summers build a cocoon in which to rest until the rains come. After a downpour, they can travel in vast concourses across the surface. Darwin noted that many species do not like to leave the doors to their burrows open, and sealed them by pulling in leaves. Others made piles of digested earth – casts – on the ground, and in some tropical forms these could be several centimetres high.

The creatures excrete with some care, for the tail, he noted, was used almost like a trowel to make a neat heap of ordure. A careful look at the body waste revealed many fine grains of silt that had been broken down from larger particles within the soil.

Worms live for no longer than two years or so, and most die younger than that. They get us all in the end, but some creatures get their retaliation in first. Many animals eat them. Badgers and hedgehogs are fond of a diet of worms, and as Alfred Russel Wallace had noticed, the natives of South America appreciated them too. In the Orinoco Basin of Venezuela smoked earthworms still form an important part of the local Indians' cuisine.

Many species regenerate their tails when cut off, and a few do the same for a head, but – in spite of myths to the contrary – none of the familiar kinds can develop into two individuals when cut into pieces (an amputated tail may grow a mirror image of itself, but then it starves). A few do reproduce by simple fission; the back breaks off and forms a new worm, and – in some – the animal splits into a dozen or more pieces, each of which gives rise to a new individual. The ability to multiply by breaking into fragments is common in the lower reaches of the animal kingdom, but the worms are the most advanced creatures to possess that talent.

The sex-life of annelids is varied indeed. Several are all-female and lay eggs without benefit of males. Some of the clonal kinds spread fast and have invaded new habitats such as sewer pipes. Others are hermaphrodites, with separate male and female genitalia. Sex happens in a long slime tube in which boy-girl meets girl-boy. The two animals lie head to tail to consummate their relationship. The male checks the virginity or otherwise of the female element of its partner and adjusts the amount of sperm to match. It increases the volume by three times when it senses that its mate has already had sex with another, no doubt to flood out the previous donation. The animals prefer to copulate underground, but sometimes move to the surface (in Darwin's words, 'their sexual passion is strong enough to overcome for a time their

dread of light'). A swollen mid-section of the body forms a protective cocoon as the eggs are laid.

Worms are among the simplest creatures to have a central nervous system, with a distinct brain connected to a set of nerve cords
(although even after the brain has been removed the animals can
mate, feed and find their way through a maze). A section of the
worm book is headed 'Mental Qualities' – but its first sentence
reads 'There is little to be said on this head.' Even so its author set
out to see just what their lowly wits were capable of. He noted
that they often pulled in leaves to seal the mouth of the burrow,
perhaps, he thought, to protect themselves from the cold. Whatever
crosses the animal's mind as it drags fronds into its home, it acts
with a degree of foresight. Worms, he found, prefer to grasp a leaf
by its tip. More than nine-tenths of broad leaves were pulled in
from that end, but for narrower kinds, which are easier to slip into
a small opening, just two in three. Those of the rhododendron
curl up when on the ground, so that some were narrower near
the base, and others near the tip. The creatures, more often than
not, pulled them in by the narrow end. They were just as smart
when it came to pine needles, which could be dragged in only
base-first.

Darwin admired such rationality, for it forged a link between the
lowest creatures and the most noble: 'one alternative alone is left,
namely, that worms, although standing low in the scale of organ-
ization, possess some degree of intelligence'. In the first real
experiment on invertebrate psychology Charles and his son
Horace presented the animals with paper triangles cut into various
shapes – and once again they acted in the most efficient way, for on
most occasions they seized the pointed end. Other studies of their
intellectual universe involved the choice of foods – meat, onions,
starch or lettuce (with beads and paper used in an attempt to trick
them). In a series of midnight expeditions to the lawns of Down
House, the father-and-son team shone lamps upon the animals,
warmed and cooled them, and subjected the unfortunate creatures

to tobacco smoke. The subjects were 'indifferent to shouts' and just as unconcerned by the shrill notes of a metal whistle or the deep tones of a bassoon. They did respond to vibration, and became agitated when placed on top of a piano. They were 'more easily excited at certain times than others', and a series of taps upon the ground made them emerge. Hungry birds could often be seen doing just that to persuade their prey to venture forth. There was, no doubt, a wide gap between their mental world and that of the naturalist – but profound as it was, it had been bridged by the same system of slow change that moulded the physical universe of each.

Those patient experiments on the inner life of the burrowers were an introduction to their wider role in the world of the soil and their ability to modify their own habitat and that of those who stride the ground above. Most of Darwin's book is devoted to the animals' impressive ability to disturb and fertilise the ground.

That talent had been noticed long before. Aristotle described the worms as the 'Earth's entrails'. Cleopatra herself decreed them to be sacred animals, and established a cadre of priests devoted to their welfare (although they were less important than scarabs, those other recyclers of dung, whose image was universal in Pharaonic times). Cleopatra's interest arose because the creatures were so important to the fertility of the mud laid down by the Nile (and they were also useful in weather forecasting). Herodotus knew as much when he wrote that 'Egypt is the gift of the Nile', and most of the immense deposit of the great river that comes down in the annual flood does indeed begin as eroded worm-casts from the Ethiopian highlands, far upstream. The same is true closer to home. In 1777, the English naturalist Gilbert White wrote, in a letter unknown to Darwin, of their 'throwing up infinite numbers of lumps of earth called worm-casts which, being their excrement, is a manure for grain and grass . . . the earth without worms would soon become cold, hard-bound and void of fermentation, and consequently sterile'. As he put it in *The Natural History of Selborne*, 'Earth-worms, though in appearance a small and despicable link in the Chain of Nature, yet, if lost, would make a lamentable chasm.'

Without their help, we would ourselves fall into a real abyss. Those simple creatures play a role in both economics and history. They improve drainage and break organic matter into fine particles: 'all the vegetable mould over the whole country has passed many times through, and will again pass many times through, the intestinal canal of worms'. That unromantic product determines the fertility of the soil, which does a lot to dictate the nature of the society that lives upon it.

In the mid-nineteenth century gardeners saw the animals as mere pests, relatives of tapeworms and such unpleasant beasts (the word vermin, indeed, has the same Latin root as does 'worm'). Books advised how to get rid of the unwelcome visitors by driving them from their burrows with mallets, poison or steel rods inserted into the ground and played with a bow (a pastime known as 'worm-grunting' to the American fishermen who still use it to collect bait). Soil was considered to be a product not of biology but of chemistry and physics, for it came from the mechanical dissolution of rock and the chemical decay of vegetation. The Comte de Buffon, mentioned in *The Origin* as a pioneer of the notion of natural selection, was, like his British successor, interested in what made the Earth's outer cloak. He noted that many soils contained grains of minerals such as iron, and that cover tends to be thinner on mountain slopes than in valley floors. All this was, he claimed, proof of the breakdown of rocks and the importance of rain, rivers and gravity in disturbing the surface. The Russians of the period – obsessed as they were with the vastness of the steppe and its effect on the Slavic psyche – were pioneers in the study of the deep and dark *chernozem*, the 'black earth' that fed the masses and nourished the nation's soul. They, too, emphasised the role of chemical decay. Why should such mundane creatures as worms play any part in the sacred soil of Mother Russia?

Physics and chemistry do, no doubt, help build the ground

beneath our feet. Chalk and limestone dissolve in the rain and even sandstone and granite can be eroded away to make earth. Tiny cracks fill with water, which shatters rock when it freezes. The surface tension that holds water to the walls of minute channels also exerts huge pressure as the liquid warms and cools. Clay itself – little more than tiny particles of ground rock – is a product of such insidious action.

The earliest fossil soil is three thousand million years old, almost as ancient as land itself. It was made with no help from biology. Three hundred and fifty million years before the present the first land plants moved on to a sterile landscape. Since then, life has fed on soil and soil on life, in a great cycle that enriches both.

The labour of the worms did a lot to improve the Down House garden. Its topmost layer is filled with channels, most of them thinner than a human hair, and around half filled with water. Below it lies a sheet of material with little air and no worms. A large part of the animals' contribution to fertility comes from their ability to open the ground to air and water. A hectare of rich and cultivated ground is riddled by ten million burrows – which, together, add up to the equivalent of a thirty-centimetre drainpipe. Half the air beneath the surface enters through burrows, and rain flows through a disturbed soil at ten times the rate of unperforated ground.

The surface of the Earth, when watched for long enough, is as unruly as the sea. Everywhere, soil is on the move. Gravity, water, frost and heat all play a part, but life disturbs its calm in many other ways. Living creatures – from bacteria to beetle larvae to badgers and to worms themselves – form and fertilise the ground. What lies beneath our feet forms the largest reservoir of diversity on the planet, with a thousand times more kinds of single-celled organism in a square metre than anywhere else. The soil contains more species than the Amazon rain forest. Its vast variety of inhabitants, large and small, burrow through the topmost layer, draw in air, digest its goodness, excrete into it and turn over so

much material that the skin of our planet is in constant eruption.
The 'biomantle', the organic layer near the surface, can be metres
deep or be no more than a thin sheet. Its base is marked by a layer
of pebbles that sinks to a depth at which the stones can no longer
be disturbed by the animals that agitate the ground above. As the
mantle churns, the relics of man's labour – from ancient tools in
Africa to the pots of the first European settlers in Australia – sink
through the topsoil, and accumulate, with the stones, just where
the tillers abandon their efforts.

Darwin's subterranean subjects have many assistants. A shovelful
of good earth contains more individuals than there are people on
the planet. Most soils have hundreds of thousands of tiny mites and
springtails in every square metre. Roots exude sugars and other
substances that feed the millions of single-celled creatures that teem
around them. They add their remains to the helpful productions of
the worms' rear ends. Bacteria and fungi possess powerful enzymes
that can break down material that even earthworms cannot digest.
They feed roots, break down vegetation – and produce antibiotics.
Until the 1930s, a diagnosis of tuberculosis was a death sentence.
The disease had killed a billion people since Darwin's birth. Then
it was found that a soil suspension attacked the bacteria responsible – and soon streptomycin was discovered and the disease was,
at least temporarily, defeated. The microbial world beneath our
feet is still almost unexplored and may have far more to offer.
Molecular probes that pick up known genes in unknown species
suggest it contains innumerable members of a very distinct group
of creatures called Archaea. They look rather like ordinary bacteria,
but in fact occupy a separate kingdom of life. Once seen as eccentric
denizens of hot springs, we now know that there may be a hundred
million of them in each gram of soil, many times more than bacteria.
Each burns up ammonia and other waste products and helps maintain the Earth's fertility.

Worms stir their habitat without cease, and as roots grow they
push barriers out of the way, and die to leave channels into which

soil may collapse. As the roots suck in water, the soil settles, and as trees lash back and forth in storms they disturb the ground. A large tree can shatter solid rock as it falls, and the hole it leaves may take centuries to fill. Small animals do even more. Insects, mites, spiders and subterranean snails, together with the worms, may make up fifteen tons of flesh in a hectare of soil – an elephant and a half's worth (and a single pachyderm needs several times that area to feed itself).

The elephant under the grass is a voracious beast. Earthworms are earth-movers, but in the tropics ants and termites may do more, for they carry up material from several metres down. Alfred Russel Wallace was astonished by the richness of the ground in some parts of Brazil: 'a layer of clay or loam, varying in thickness from a few feet to one hundred . . . over vast tracts of country, including the steep slopes and summits . . . of a red colour, and is evidently formed of the materials of the adjacent and underlying rocks, but ground up and thoroughly mixed'. It had been mixed by ants.

Larger creatures also help to stir up the soil, and elephants themselves often paw away at the surface. Below ground, moles, prairie dogs, marmots, wombats, meerkats, badgers and other excavators join in, each in its own part of the world. They are helped by aardvarks, armadillos and anteaters as they scratch away in search of food. A colony of naked mole rats can build burrows a kilometre long. In the south-western United States, tens of thousands of symmetrical piles of earth up to two metres high and fifty metres across mystified historians for years. They were, they imagined, sacred sites of a lost tribe of Indians. The truth about the Mima Mounds is more prosaic. They are built by gophers, which over the millennia push tons of earth uphill to provide a dry refuge in a marshy place. Even the bottom of the sea is not safe, for manatees and narwhals dig up food, skates do the same and shrimps work away at the top few centimetres of mud. Enthusiasts for the process trace it back to the Cambrian explosion, around five hundred and forty million years ago, when the first animals with hard shells emerged. They were able to dig into the thick layered mat of microbes that had until then covered the seabed. As they did, a whole new way of existence sprang into being. The revolution of the burrowers marked the origin of modern life, and their descendants are still essential to keep it healthy.

Today's worms are merchants as well as miners, for they are major players in the vast traffic in chemicals that passes from the world of life to that of death and back again. Darwin knew as much when he wrote that 'All the fertile areas of this planet have at least once passed through the bodies of earthworms.' In an English apple orchard they eat almost every leaf that falls – two tons in every hectare each year, and in the same area of pasture certain kinds munch their way through an annual thirty tons of cow dung. A few tropical species pile their casts into mounds twenty centimetres high and his book refers to the gigantic castings on the Nilgiri Hills of southern India as an indication of the vast amount the animals must chew. Most of their endless meal is ground down in the muscular gizzard. Rather little is absorbed. Even so, it undergoes chemical changes. The experimenter fed some of his subjects with soil laced with red iron oxide powder and noted that it lost its colour when excreted; proof that acid and enzymes had done the job. Their potent guts change the soil, for the chemistry of clay is much modified when passed through their bodies. It is ground even finer than before, which helps it retain water and nutriments – and the tiny particles left after the worms have done their work mean that clay has ten thousand times the surface area of an equivalent volume of sand.

Some species of worm have the unexpected ability to draw – like certain plants – carbon from the air and convert it into soluble substances that can be recycled. *Vegetable Mould* suggests that the small grains of chalk found in the digestive glands are waste products. The truth is more remarkable. Radioactive labels show that the glands extract carbon from free carbon dioxide –

abundant beneath the soil – at a considerable rate (an unusual talent for an animal) and combine it with salts of calcium. The particles of chalk so produced are excreted and also return to the earth when the creature dies. The worms hence do a lot to increase soil carbon and to improve fertility.

The constant flood of slime pumped out as they burrow recycles other minerals such as nitrogen. Plants and animals die, and farmers pour fertilisers, manure and treated sewage on to their lands and the worms do their bit to pull them into the earth. Their casts contain five times as much nitrogen and ten times as much potassium as does the soil itself. A large part of that emerges from their busy inner life; to the bacteria that live in the oxygen-free world of the gut. Each worm intestine is a tiny fermentation chamber in which bacteria chew up manure. They make useful fertiliser – but with the side-effect that they also pump out nitrous oxide, a greenhouse gas (and, as 'laughing gas', a primitive anaesthetic), which gives their hosts an unexpected role in global warming.

A simple experiment shows the power of the worm to disturb the underground world. A mouse carcass was placed in a glass jar with some fine rubble and leaves, plus an added earthworm. In just three months, the bones had been scattered sideways across about ten centimetres and some had been dragged the same depth into the soil. In wormless jars, the corpse stayed undisturbed. Darwin, too, set out to test his subjects' powers of burial. On morning after morning, in the garden at Down House, he counted the number and size of casts – each the undigested remains of a worm's meal – and found dozens in a typical square yard. His cousin Francis Galton joined in and, ever keen to use statistics, counted the number of dead worms he saw on paths in Hyde Park. He found, on the average, a corpse every two and a half paces. The worms, he calculated, brought seven to twenty tons of earth to the surface in every acre of his local fields each year. At that rate, worms would lay down half a centimetre of top-soil in a twelvemonth. In fact, their labours are even more impressive, for most of what they excrete remained beneath the surface, invisible to the eye.

The number of worms is so huge, and their labours so sustained, that in time they can do great things. In a follow-up of his youthful observation at Maer, and soon after moving to his own grand house, Charles Darwin scattered quantities of broken chalk and brick over a field near Downe to test how fast it sank. Twenty-nine years later he dug a trench across the chalk site, and found most of the chalk buried some fifteen centimetres down. The bricks, on thinner soil, took longer but even they disappeared in the end. By 2005, the fragments of brick had sunk to the level of a solid band of flinty clay into which the worms could not penetrate, while the chalk had been dissolved away.

Darwin's garden had ten or more burrows in every square metre. Given the ability of each animal to chew through earth, if they acted with equal enthusiasm in every cubic centimetre the whole mass would be disturbed to a depth of a metre or so in about five thousand years. That was not at all the case, for stone tools of that age are often found at shallower levels. In addition, many species of worms reuse their burrows and that economical habit also reduces the extent to which they agitate the ground. As a result, an object that falls on the surface may sink quite fast in its first few decades, but then slow down.

In his final decade, Darwin started an experiment to test their sepulchral power. He placed a lump of rock – a hefty millstone forty centimetres across – in a corner of his lawn. A long brass rod was pushed deep into the soil through a hole in the centre. The movement of the rock in relation to the rod measured the efforts of the burrowers as they worked away below. In its first days, it sank by around twenty millimetres a year. Charles died before the experiment was complete, but his son Horace continued the study and found that the worm-stone sank by twenty centimetres in ten years. Today's stone, admired by the curious as it might be, is a copy of the original and has been moved since it was first put in place. Nowadays it sinks more slowly than it did. Sir Arthur Keith (who became wrapped up in the Piltdown Man scandal before writing an early biography of Charles Darwin) retired to live close to Down House in the 1930s, and re-examined the sites used in the chalk and brick experiments. Eighty years on, the marked stones had sunk little more than they had in the lifetime of those who set them there, as further proof that the worms are most active near the surface.

At Down House, the longest-running biological experiment in the world is still under way but, ancient as it might appear, the worm-stone has been in place for no more than an instant of geological history. Darwin realised that in the abyss of time his own life and the span of his own experiments were fleeting indeed. He saw that the remnants of ancient structures scattered over England gave him a better chance to test his subjects' powers. In late middle age, he began a tour of the stately ruins of England and – ever a busy correspondent – wrote to dozens of people who might give him information.

The head of excavations at Wroxeter, near Shrewsbury, came up with a strong hint of what worms could do, given time. The city had been founded by the Romans to act as the capital of a British tribe, the Cornovii. Viroconium, as it was called, at its peak held six thousand people. In time, it fell into decay and became, in legend, the site of King Arthur's court. Camelot, the archaeologist responded, was in some places buried under more than a metre of vegetable mould. Much of that was due, Darwin had no doubt, to the efforts of earthworms.

In 1877, men at work on the restoration of Abinger Hall in Surrey, the grand house of his friend Thomas Henry Farrer, who had earlier helped with the experiments on hops and other climbers, discovered the remains of a Roman villa. The Sage of Downe came to visit. He saw how the creatures crawled through the rotten concrete floor of the ancient structure, and brought up material from below. At the time, and for long afterwards, antiquarians assumed that the layers of earth found above decorated pavements and the like were the remnants of later and less civilised inhabitants, who had settled down in the houses of their erstwhile masters and left their household rubbish behind. The supposed squatters were, in truth, worms.

Darwin was impressed to discover burrows almost two metres beneath the modern surface. The animals could even mine into the ancient structure's thick walls. Farrer observed their activities for several weeks, and saw them hard at work as they heaved the soil. A quick sum showed that their labours were more than sufficient to bury a Roman house within a few centuries. At a villa with a mosaic floor on the Isle of Wight, Darwin's son William was told that so many castings were thrown up between the tiles that the ground had to be swept every day to keep the pattern in view. William also visited Beaulieu Abbey in Hampshire and found that as a result of their labours the bottom of a hole dug down to the ancient floor twenty years before was already covered.

The trip to Stonehenge was also part of the worm project. It showed that, active as the animals might be in rich soils, in some places they achieved rather less. Emma herself noted that they 'seem to be very idle' and in that thin soil the animals had done no more than enough to sink some of the 'Druidical' stones by twenty centimetres or so since they had toppled (they rested on the chalk layer beneath, into which the creatures could not penetrate). John Lubbock, who lived close to Down House, had dated the stones to the Bronze Age, which began in Britain around 2100 BC. The latest estimates push their masons further into antiquity at close to 2300 BC – a period when the Britons began to cut down their forests and replace them with fields. Some of the monoliths fell long ago, in part through the efforts of the worms themselves, whose work, and that of the rain, weathered away the soil that once supported them. Others fell – or were pulled down – within the past few centuries (one major collapse happened in 1797), which suggests that perhaps the burrowers were not as idle as Emma imagined. Indeed, they buried the stone chips left by the first modern excavators of the site in the 1920s to a depth of about five centimetres or so in thirty years, which was almost the same rate as that measured at Down House.

In Charles Darwin's sesquicentennial year of 1959 a plan was hatched for an improved version of his experimental millstone, built on a grander scale, to test the destructive effects of such creatures on the ancient monuments of England. The British Association – that stamping ground of Victorian evolutionists – set up a Committee to Investigate by Experiment the Denudation and Burial of Archaeological Structures. A long pile of chalk, with a ditch alongside, of about the size and shape of a typical section of an English barrow or burial mound of three thousand years ago, was built at Overton Down, not far from Stonehenge itself. Plant spores and bits of broken flowerpot were scattered on the surface. Just thirty years later, natural weathering and the efforts of Darwin's favourite excavators had caused large parts of the wall to collapse into the ditch, and both were covered with a layer of grass and soil. The pieces of broken pottery moved by around three centimetres a decade, and the spores were carried several centimetres into the depths. The worms were at work; and a similar structure built at about the same time on an acid heath in Dorset, with far fewer of those animals, was far less disturbed. At Overton, the experimental barrow now looks much like others a hundred times older. Once again, most of the change in the soil took place in the first few years after it had been disturbed. The next survey is planned for 2024, when, no doubt, the British Association Barrow will be almost impossible to tell from those built by the associated British long before.

Life's underground frenzy soon blurs the record of the past. At Abinger Hall, several Roman coins were found; but among them was a halfpenny dated 1715. An incautious student would gain an odd view of British history if he took that observation literally. In a five-thousand-year-old Indian mound in Kentucky, the constant activity of soil animals has been enough to turn over and mix up

the whole of the site fifty times over since the original inhabitants
left. Sites with moist, rich soils are at more risk of disturbance than
are deserts or cold uplands – but as men and worms have similar
tastes in places to live, the news for those who hope to reconstruct
ancient history is bad.

On Leith Hill, the highest in south-east England, Darwin
tried to test the extent to which the material dug up would slide
downwards to fill valleys and plains. He found that the castings
soon rolled downhill and reckoned that for a steep slope a hundred metres long ten kilograms of earth would be washed to the
bottom each year. His estimate is close to those made today, and
is a tribute to the worms' importance as architects of the fertile
fields of southern England – and of the hungry pastures on the
hills above. The wind, too, can transport their excreta, to add
another weapon to the animals' armoury as soil engineers. Wind-
blown soils make up large parts of China, the Great Plains and
the Rhine Valley. A strong gale moves stones and gravel, but such
large elements soon fall to earth and the finest, and most nutrient-
rich, particles – those of the worm-casts included – are blown
furthest of all. That valuable powder can even cross the Atlantic.
On the last leg of the *Beagle* voyage, the young naturalist noted a
fall of white dust on to the deck of the ship as it sailed off South
America. Some of that came from North Africa, thousands of
kilometres away. Silt around Lake Chad – in part the product of
worms and their fellows – is picked up by gales, and sifted finer
and finer as it travels, until it becomes filled with valuable salts of
nitrogen and phosphorus. More than ten million tons of the
stuff fall on the Amazon rain forest each year and bring fertility
to those thin and hungry lands. The good work of the worms
can, it appears, cross great oceans.

Nowhere is their power better seen than when they themselves traverse the seas. Some species – the 'peregrines' – are keen
migrants. In New Zealand at the end of the nineteenth century,
farmers found to their surprise that what had been thin pasture

had been transformed into lush loam. The immigrants were at work as they broke down soil into compost. They can move into empty pastures at ten metres per year. In today's New Zealand, as they continue to spread, they can bury metal rings – a modern version of the worm-stone – at twice the speed measured in the Down House garden. Now, the Europeans are on every continent apart from Antarctica and in many places far outnumber the natives.

Their ability to improve the ground is so impressive that the animals are sometimes introduced to heal the damaged earth. After mining is finished, or all the peat has been stripped from a bogland, the intruders do a lot to help a landscape to recover. In the Kyzylkum Desert of Kazakhstan and Uzbekistan vast numbers were moved in Soviet times to isolated oases, with salutary effects. Waterlogged Dutch polders, too, had their drainage improved by a hundred times after the animals were called in to help the engineers who had recovered the fields from the sea.

Long before they began to move, the worms helped make the landscapes of the agricultural regions of the world (southern England included), and as an incidental maintained innumerable farms and gardens in a fertile and healthy state. Now, those who till the ground – like their predecessors when farming began, but at a far greater rate – are undoing the animals' work. Like skin stretched too tight on an ageing face, the Earth's epidermis – the soil – has grown thinner with the years. Like age itself, the process is slow but impossible to resist, and like the signs of decay in a human body, the process speeds up with the years.

Man has flayed his home planet for ten millennia. Soil is hard to make, but easy to destroy. A modern plough shifts hundreds of tons a day, which is beyond the capacity of the most vigorous invertebrate. It digs down to no more than a metre or so, to make a solid and impermeable layer at just the depth of the blades. Another problem arises when fifteen-ton tractors roll across the surface. Their wheels compact the loose soil into a material almost like concrete, in which nothing will grow. In addition, continued ploughing breaks up the topmost layer and allows vast quantities to wash away. Every farm's raw material is on the move, from hill to plain, from plain to river and from land to sea. The evidence is everywhere. My parents' house overlooked the Dee Estuary (the Welsh rather than Scottish version). What was, a few centuries ago, a broad waterway has become a green field with a ditch in it and the local council is exercised about what to do about the sand that blows on to its roads. The reason lies in the fertile fields of Cheshire and North Wales. They have been ploughed again and again and their goodness has disappeared downstream.

The process is speeding up. The amount of organic carbon in Britain's lakes and streams has rocketed in the past twenty years and in some places has almost doubled. Waters that once ran clear now flow with the colour of whisky, which itself gains its hue from the carbon-rich streams that run through peat bogs to feed the stills. On the global scale, matters are even worse. Twenty-four billion tons of the planet's skin are washed away each year – four tons for every man and woman – and although some is replaced and some has always been lost to the rain and to gravity, the figure is far higher than once it was.

Man has long been careless of the deposits in his soil bank. Again and again, as a civilisation grows it empties its underground accounts, goes into decline and collapses. Usually it takes around a thousand years. Marx himself noticed, for he wrote: 'Capitalistic agriculture is a progress in the art, not only of robbing the worker, but of robbing the soil.'

Darwin compared the work of the worms with that of the plough. Since his day, farm machines have become far more powerful. His experiments showed how earth could slip and churn, but he had no more than primitive tools to measure how much movement there now is. A sinister spin-off of modern technology has come to the aid of science. From the first atom bomb in Nevada in 1945 to the last air-burst of a hydrogen bomb in 1968, vast quantities of radioactive fallout spilled across the world. Radioactive caesium, which has a half-life of around thirty years, binds to soil particles. In an undisturbed site, the element is most abundant near the surface, but after a plough has passed the radioactivity is dispersed to the depth of its blade. As the disturbed ground is washed away, the irradiated soil is lost, to accumulate in places where the mud settles. In the Quantock Hills of Somerset, soil is, in this era of industrial agriculture, lost at around a millimetre every year. As the land sinks, the ploughs dig deeper, and the relics of the Romans that lie beneath will be smashed within a century – although, thanks to the worms, they have been preserved for two thousand years. Already, most of the treasures picked up by metal detectors come from ploughed fields, proof of how fast man's machines are stripping the fragile surface of the globe.

The Romans themselves paid the price for their abuse of the soil, for at the time of their decline and fall so much damage had been done to Italy's farmlands that a large part of the Empire's food had to be imported. The fertile fields around their capital lost their goodness and vast quantities of grain were shipped in from Libya, which in turn became a wasteland as its surface was stripped by ploughs. In Rome's last Imperial years, it took ten times more Italian ground to feed a single citizen than it had during its heyday.

The damage had begun long before. The first large towns appeared in the Middle East around eight thousand years ago. Quite soon their growing populations began to demand more food. The farmers exploited their precious mould with no thought of replacing its goodness or allowing their fields to rest. Instead they attacked it with ceaseless vigour. The plough was invented soon after oxen were domesticated. In a few centuries the topsoil was gone and many villages were abandoned. Within a couple of millennia, all the fertile land of Mesopotamia was under cultivation. Many irrigation canals were built. The soil was soon washed away, and the canals became blocked with mud. Enslaved peoples such as the Israelites were forced to clear it (and the abandoned city of Babylon is still surrounded by dykes of earth ten metres high, the remnants of their labours). Abraham's birthplace, Ur of the Chaldees, once a port, is now nearly two hundred and fifty kilometres from the sea, and the plain upon which it sits is the remnants of what were fine fields, lost to leave a desert. Salt poisoned the last of the land, and what had been the Fertile Crescent – and the civilisation it fed – collapsed.

The same happened in China, where the Great River was renamed the Yellow River two thousand years ago as swirling earth changed the colour of the water. The ancient Greeks, too, faced the problem when the country around Athens was stripped bare. Plato blamed the farmers.

The real disaster for the skin of the Earth came in the Americas. The Maya bled their landscape dry, as did the inhabitants of Chaco Canyon and Mesa Verde – the abandoned Native American settlements that now form part of the deserts of the American south-west. The Europeans were even worse. Virginia's 'lusty soyle' was ideal for tobacco, but the plant sucks goodness from the ground as much as it does from the bodies of those who consume it. In three or four years of that hungry crop the soil was drained of goodness, but farmers saw no need for fertiliser ('They take but little Care to recruit the old Fields with Dung') and simply moved on to the next piece of land. George Washington himself complained that exhaustion of the soil would drive the Americans west, as it soon did. Charles Lyell, Darwin's geological mentor, used the huge gullies that scarred the devastated surface of Alabama and Georgia to examine the rocks below and commented that soon American agriculture would collapse.

The nemesis for vegetable mould – and a major threat to our own future – began in 1838 when John Deere invented the polished steel plough – 'The Plow that Broke the Plains', as the

memorial at his childhood home calls it. Soon thousands of his
devices were tearing up the prairies.

Since farmers began to work the Great Plains, the soil has lost
half its organic matter. The Mississippi is what Mark Twain called
America's 'Great Sewer', and the amount of silt that pours down
it has doubled since John Deere's day. As the crew of the *Beagle*
had noticed, yet more is taken by the wind. A thin layer of sod
had kept the prairie soil in place and soon it began to blow away.
The Dust Bowl followed. A huge gale in May 1934 blew a third
of a billion tons of dust eastwards from Montana, Wyoming and
the Dakotas. The dense cloud reached New York two days later,
and petered out far out in the Atlantic. The finest material was
blown furthest, which meant that the New Englanders gained the
nutriment-filled dust that had passed through the guts of Montana
worms, while the unfortunate westerners were left with just a
rough and hungry silt. The gales returned again and again until by
the mid-1930s more than a million hectares of prairie had been
replaced by desert.

The situation elsewhere is worse. Worldwide, an area larger
than the United States and Canada combined has already been
despoiled. In Haiti, almost all the forest has gone and thousands of
hectares of ground are now bare rock. Less than a quarter of the
island's rice, its staple food, is now home-grown and over the past
decade food production per head has gone down by a third. In
China, the Great Leap Forward exhorted the peasants to 'Destroy
Forests, Open Wastelands!' They did, and the soil paid the price.
A parallel problem in Africa explains some of the continent's
chronic instability. Across that landmass, three-quarters of the
usable land has been bled of its nutriment by farmers, who cannot
afford fertiliser and whose fields are, as a result, no more than a
third as productive as those elsewhere. Its earth still leaches its
goodness into the water, or into dust. The Sahel, the area of thin
soil to the south of the Sahara, is becoming a dust bowl and loses
two centimetres of surface each year. Hundreds of millions of

people go hungry as a result. A planet ploughed by man is far less sustainable than it was when tilled by Nature.

In 1937, after the Dust Bowl disaster, President Franklin D. Roosevelt said in a letter to state governors that 'The nation that destroys its soil destroys itself.' His Universal Soil Conservation Law became the first step to putting right the damage done to the precious fabric of his nation. It promoted careful ploughing, the use of windbreaks and a ban on the reckless destruction of forests. Within a few years, most of the American Dustbowl returned to a semblance of health. To plough with the contours, rather than against them, makes a real difference (and the Phoenicians had the same idea). In China, too, the Three Norths project plans a five-thousand-kilometre strip of trees in an attempt to stop the light earth from being taken by the wind. Even the Sahel has gained hope from low technology, with lines of stones set across the slopes to stop the thin earth from washing away. In Niger alone, fifty thousand square kilometres of land have been put back into cultivation.

Soil protection of this kind has a long and unexpected history. In most of the Amazon basin, the soil is bitter and thin for constant rain leaches goodness away, and the vegetation feeds on itself as it recycles nutrition from its own dead logs (which is why cleared forests are so infertile). Patches of the so-called Terra Preta ('black soil') lands, in contrast, are small islands of deep dark earth scattered among those starved soils. They were formed not long after the birth of Christ by native peoples who settled down, fed slow fires with rubbish and leaves, piled up excrement and added bones to the mix. They lived in large and scattered cities that recycled their rubbish and set up, in effect, precursors of the green belt that surrounds many modern towns. Over the centuries carbon levels shot up and the ground became filled with worms and their helpful friends. They chew up the ashes and excrete it as a muddy paste of carbon mixed with mucus. So fertile is terra preta, with ten times the average amount of nitrogen and phosphorus, that it is now sold to gardeners.

In Amazonia, the United States and elsewhere, the worm has begun to turn, although perhaps too late. The new concern for the soil is manifest in many ways. Organic gardeners use 'vermicompost' – the end-product of species such as the red wriggler fed with waste from farms or factories – as a powerful boost to the garden. Real enthusiasts make their own in bins into which they throw household rubbish and old newspapers to be transformed into fertiliser. They are part of a global movement which sees commercial agriculture of the kind that destroyed the prairies as an enemy and tries, in a small way, to replace what has been lost.

On the larger scale, too, the ploughman is going out of fashion and the burrowers and their fellows are allowed to work undisturbed. Farmers now scatter seed on undisturbed ground, or insert it through the previous year's stubble. The world has a hundred million hectares of such 'no-till' agriculture, and even in places where the plough still rules, the land is treated with more care than before. Brazil, in particular, with its deep ant-built landscapes, has been a pioneer. No-till farming does its best to leave the hard work to Nature. Instead of clearing the ground of dried stalks or leaves, the remains of a crop are left behind. They are soon dragged underground. As a result the earth can take up more carbon and become more fertile. Weeds are suppressed by the mat of dead vegetation that appears, water runs away more slowly and the temperature of the surface is less variable, and better for planted seeds, than in a bare ploughed field. Farmers who once saw worms as a pest now realise that to let them flourish undisturbed does more to preserve the ground – and to make profits – than to use machines.

The move away from technology and the return to Nature's tillers has been a real success. In some places it reduces soil loss by fifty times, and the habit is spreading fast. In Canada, two-thirds of the crops now grow on earth left unploughed, or ploughed in such a way as to reduce the damage. An era in which millions of tons of mud are lost from fields may be succeeded by an age in which plants and animals – their efforts feeble as individuals, but all-powerful en masse – are allowed to regenerate the soil. Darwin, with his passion for the natural world, would be pleased; but there is still a lot to be done. The risk of a global Mesopotamia – a collapse in food production as the worms' precious but forgotten products are squandered – has not gone away.

The great naturalist himself, in old age, often spoke of joining his favourite animals in 'the sweetest place on earth', the graveyard at Downe. He was denied the chance to offer himself to their mercies for he was interred in Westminster Abbey, whose foundations kept them out. His remains may have been saved from the annelids but have no doubt been consumed by other creatures. Darwin's simple idea – of the importance of gradual change in forming the Earth and all that lives upon it – has replaced the power of belief with that of science. Perhaps today's science of the soil will return the compliment by allowing worms and their fellows to restore the damage done by his descendants to their most important product: the vegetable mould that keeps us all alive.

ENVOI

THE DARWIN ARCHIPELAGO

When Charles Darwin moved to Down House in 1842 the population of England was fifteen million and London, the largest city in the world, had some two million inhabitants. At the time of his death, forty years later, the number of Londoners had doubled and the capital's fringes were creeping towards his retreat. The city has multiplied by four times since then and his home has become a museum. It sits in a well-preserved enclave that pretends to be a village but is in fact part of the London Borough of Bromley. A few segments of countryside nearby have been preserved (and the 'tangled bank', the microcosm of life referred to at the end of *The Origin*, is almost unaltered) but the landscape around Down House has become suburban at best. A glance across the famous Sand-walk reveals the tailfins of planes parked a few hundred metres away. Biggin Hill Airport was a Battle of Britain fighter base whose pilots claimed to have shot down more than a thousand aircraft.

Now the place is the most popular light aviation centre in England and its annual air fair attracts a hundred thousand visitors. A lot has changed in the Kentish countryside since Charles Darwin walked across what has become an oil-stained strip of tarmac.

Many other places associated with the great man – and many of his subjects, from apes to earthworms and from insectivorous plants to *Homo sapiens* himself – have been transformed since his demise. That would be a surprise to the patriarch of Down House. Darwin looked at the past to understand the present. He scarcely considered what the future might bring, for in his view evolution was so slow, and flesh so stable, that no real changes in the world of life were to be expected for many generations to come. In the long term, no doubt, the outlook was bleak: as he wrote in a letter to his old friend Joseph Hooker: 'I quite agree how humiliating the slow progress of man is, but every one has his own pet horror, and this slow progress and even personal annihilation sinks in my mind into insignificance compared with the idea or rather I presume certainty of the sun some day cooling and we all freezing. To think of the progress of millions of years, with every continent swarming with good and enlightened men, all ending in this, and with probably no fresh start until this our planetary system has again been converted into red-hot gas. *Sic transit gloria mundi*, with a vengeance.'

A moment of hindsight on his two hundredth birthday shows that he was badly out in his timing of the coming apocalypse, at least when it came to biology. Every continent is indeed swarming with men, but 'progress' (if such it is) has not been slow, but meteoric. A great deal has happened in the evolutionary instant since 1809. In the next two centuries, plants, animals and people will see an upheaval greater than anything experienced for thousands of years. For most plants and animals the prospect of biological annihilation is far closer than the certainty of the heat death of the universe.

In the last few weeks of his voyage, in July 1836, the young explorer had a brief vision of what lay ahead. The *Beagle* dropped anchor at St Helena, halfway between Africa and South America. The island, first occupied by the Portuguese in the sixteenth century, is among the most isolated places in the world. Darwin was delighted by it: a hundred square kilometres of volcanic mountain rose 'like a huge black castle from the ocean'. He admired the 'English, or rather Welsh, character of the scenery' and noted to his surprise that St Helena's vegetation, too, had a British air, with gorse, blackberries, willows and other imports, supplemented by a variety of species from Australia. Over seven hundred plants had been described – but nine out of ten were invaders. They had driven the original inhabitants to extinction or to refuges high in the mountains. A sweep of thin pasture near the coast was known to locals as the 'Great Wood' – which is what it had been until a hundred years before, when the trees were felled and herds of goats and hogs consumed their seedlings and killed the forest. Plagues of rats and cats had come and gone as they ate themselves to extinction. On his first day ashore, he found the dead shells of nine species of 'land-shells of a very peculiar form' (one of the few mentions of snails in his entire oeuvre) and – in an early hint of evolution – noted that specimens of a certain species 'differ as a marked variety' from others of the same species picked up a few kilometres away. All those molluscs apart from one had been wiped out and replaced by the common brown snail of English gardens.

Two centuries or thereabouts after his visit, life on St Helena is worse. The island once had forty-nine unique species of flowering plant and thirteen of fern. Seven have been driven to destruction since the arrival of the Portuguese, two survive in cultivation and many more are on the edge. The last St Helena Olive died of mould in 1994. Parts of the tree-fern forest of the high mountains – still in robust health at the time of the *Beagle* – remain, but other unique habitats visited by Darwin, such as the dry gumwood, have gone and of the ebony thickets just two bushes remain. The island's giant earwig (at eight centimetres the world's largest), its giant ground beetle and the St Helena dragonfly, all common in the 1830s, have not been seen for years and Darwin's snail of peculiar form is now reduced to a population of no more than a few hundred. The St Helena Petrel is extinct and a solitary endemic feathered creature, the Wire Bird, is left. That too is under threat.

Three months after the farewell to St Helena, the naturalist's diary records that 'we made the shores of England; and at Falmouth I left the *Beagle*, having lived on board the good little vessel nearly five years'. His account of the expedition ends with a spirited enjoinder to all naturalists 'to take all chances and to start on travels by land if possible, if otherwise on a long voyage'. Charles Darwin never left British shores again.

He had no need to, for, as this book has shown, the plain landscapes of his own country gave him the raw material needed for a life filled with science. Darwin's fifty years of work on his homeland's worms, hops, dogs and barnacles changed biology for ever. Since then the British Isles have provided another useful lesson for students of wild Nature, for their modest archipelago is a microcosm of the global upheavals that have taken place since the naturalist came home.

Bartholomew's *Gazetteer of the British Isles*, in its edition published just after the great naturalist's death in 1882, describes Kent, 'the Garden of England', as a paradise: 'The soil is varied and highly cultivated . . . All classes of cereals and root produce are abundant, as is also fruit of choice quality and more hops are grown in Kent than in all the rest of England. The woods are extensive . . . Fishing is extensively prosecuted . . . of which the oyster beds are especially famous.'

A lot has changed since then. The local farms bring in half what they did even a decade ago. The oysters are almost gone and the salmon fishery of the Thames Estuary, which fed the apprentices of London with such abundance that they refused to eat fish more than once a week, has collapsed. Bucolic pursuits have been replaced by that invaluable product, 'services', which accounts for three-quarters of the county's contribution to the nation's wealth. Kent is a dormitory of London and London has become a staging post for the world. The flow of people, power and cash has carved up its landscape with motorways, rail links and webs of power lines. The oast houses that once stored hops have become commuter homes and the hops themselves – the raw material of so many experiments – cover a fraction of the fields he knew. So far, the Great Wen has been kept in part at bay by the Green Belt, but plans for a 'Thames Gateway' mean that yet more of the Garden of England will soon be a bland suburb.

Much of Charles Darwin's work on insectivorous plants, on self-fertilisation and on orchids took place in Ashdown Forest in the adjacent county of Sussex, where his cousin Sarah Wedgwood had a house and where he often walked, mused and botanised. It shows how fast the wild can retreat. In his day the forest was just one of several vast belts of English heath, successors of ancient tracts of trees felled thousands of years ago (Cobbett saw its thin soils as 'the most villainously ugly spot I ever saw'). Ashdown Forest was used by the Normans as a game preserve and was closed off with a thirty-kilometre bank. In time most of the trees were cleared and burned, in iron foundries as much as domestic hearths, and it became heathland, a semi-natural part of the semi-natural landscape that is England. Since the date of publication of *The Origin*, nine-tenths of the nation's heaths have been lost. The forest, at two and a half thousand hectares, is the largest piece left but even that is a shadow of what it was. The acid grass and marshes have been taken over by bracken or have dried out as water is pumped away to slake the thirst of millions. Many once familiar species – gentians, asphodels, sundews, orchids and more – are rare where once they were abundant and some of his favourite walks have become suburbs, farms or golf courses.

Ashdown Forest is a microcosm of the modern age. The Common Plants Survey keeps count of sixty-five of Britain's most abundant flowers, from primroses to bluebells and foxgloves, in five hundred random plots scattered across the nation. A century ago, those species were almost everywhere. By 2007, a quarter of the study sites had none of them at all. Most of the empty plots were in huge fields of corn or on wide pastures without hedgerows. Others were in woodlands. England's forests – preserved by the Woodland Trust, the Royal Society for the Protection of Birds and the National Trust as they may be – have lost large parts of their diversity. A stable ecology maintained by the labours of woodmen has been replaced by museums of elderly trees in which bluebells and foxgloves, sparrows, cuckoos and jackdaws are in decline. Across England – and across Europe – the fields are even more starved of life. Subsidies have made a desert and called it farming.

Kent's sorry tale is repeated in many of Darwin's favourite British places, from Stonehenge to Shrewsbury and from Wales to the Highlands. In a further blow to the products of evolution, the world has come to Kent and the animals – and people – of Kent have migrated to the world. Darwin's archipelago has been united with the globe, which has become a single giant continent rather than a series of islands, real or metaphorical. For his own county the Channel Tunnel makes that reality. Humankind, too, has been homogenised, for even genteel Bromley now has a tenth of its citizens from ethnic minorities. The struggle to exist for both man and beasts has become a worldwide conflict rather than a series of local skirmishes. No longer does evolution mould the natives of each corner of the planet to fit their own domain. Some creatures thrive in the international arena; but many more are doomed.

Evolution generates difference. One species and one alone has put the process into reverse. Man has instituted a simplification almost as grand as that brought by the catastrophe that destroyed the dinosaurs. The Galapagos themselves are a stark reminder of what he has done in less than two centuries. HMS *Beagle* visited the island of James in 1835. Food was plentiful: 'We lived entirely on tortoise meat . . . the young tortoises make excellent soup.' In those inelegant creatures, Darwin saw, without realising it, his first hint of evolution, for the animals from that island were distinct from those on Indefatigable and Albemarle nearby. In a rare conjunction of taxonomy with gastronomy, he noted that the James specimens were 'rounder, blacker and had a better taste when cooked' – which at the time seemed little more than a curiosity but was in fact an introduction to the biology of change.

Now, the tortoises of James and its fellows have been driven almost to extinction. From a quarter of a million in the *Beagle*'s day, their numbers have dropped to fifteen thousand. Three of the fourteen unique races have gone and a solitary animal, the famous Lonesome George, is left from another (now, at the age of ninety or so, he has been persuaded to mate with a female of a different race in the hope of preserving his genes). Pigs, as much as men, have done the job, for they love to feast on tortoise eggs. Less obvious pests have also made their way to the archipelago. The cotton cushiony scale insect invaded twenty years ago. It has reached across the whole archipelago and attacks dozens of kinds of native plants.

Pigs and scale insects are dangerous because they have Epicurean tastes. They are happy to try anything once and – like the young explorer with the tortoises – will try a novel source of food if their usual diet is not available. They can, as a result, snack on the last specimens of an endangered species without eating themselves out of house and home. Such generalised predators, as they are called, are a real threat to diversity. On the Galapagos, goats and cats are a plague, pigeons have pushed out their feathered relatives and alien wasps have done terrible damage to the insects. The islands face an era in which specialists, evolved to fit their own small place in nature, have fallen to loutish strangers able to cope more or less anywhere. A tourist on the Galapagos today – and a hundred thousand arrive each year – has less to admire than did the crew of the *Beagle*. Next century's visitors will find the place more or less indistinguishable from South America for many of its natives will be gone. The products of millions of years of isolation have been destroyed by man, the most generalised predator of all.

The Galapagos are the icons of evolution and their problems get plenty of attention. Many other oceanic islets across the globe – rare, specialised and fragile as their natives are – face the same cataclysm or worse, but not many people notice. From St Helena to Tahiti and from Hawaii to the Cape Verdes, the alarm has at last been raised. It is too late to save the majority of such places, most of which began their decline long before the *Beagle* arrived.

The fate of the giant earwig of St Helena or the tortoises of the Galapagos is sad enough but Charles Darwin's less spectacular subjects provide a more trenchant statement of the universal attack on the biosphere. They are both under threat and a threat to other places. The humble creatures he studied – the earthworms and bees, the primroses and orchids, the plants that climb and those that snap shut on their prey – all face an ecological earthquake, wherever they may live. In many ways the lessons to be learned from such modest beings are more alarming than are those from the spectacular inhabitants of distant Pacific isles. The crisis has moved well beyond the exotic, and what was once common, or even commonplace, has become rare.

The sundews of Kent and Sussex are far from safe and many of the insectivores sent to Down House from across the globe are in deeper trouble. The wide fields of Venus flytraps and of pitcher plants that once covered parts of North and South Carolina have been destroyed. Agriculture and drainage tear up their homes and the gardeners who dig them up do not help. A more subtle threat comes from fire control, because such beings thrive best in places often burned – which in today's carefully managed countryside happens less than once it did.

Darwin's other subjects, the orchids, face the same problems. Their enemies are those of the insectivores: aggressive farmers, fragile habitats and greedy collectors. A third of the fifty British species are under threat, and several have populations of fewer than a hundred individuals – and one, the Lady's Slipper Orchid, was for a time reduced to a single plant in the Yorkshire Wolds (thousands of greenhouse specimens have now been sown in the hope that the species can be rescued). The Victorians suffered from 'Orchidelirium', and paid large sums for rare specimens. Traders destroyed whole beds to ensure that their own stock kept their price and Darwin's colleague the botanist Joseph Hooker noted how the area around Rio de Janeiro, visited by the young naturalist thirty years earlier, had even by then been pillaged of its orchids, which never reappeared. Unlike the Dutch tulip fever of the seventeenth century, which faded away, the orchid mania is still upon us, with a global trade worth ten billion dollars a year. Expensive specimens sell for thousands. Some of the business is legitimate and the plants are cultivated or cloned in huge numbers from cells taken from one or a few individuals. Plenty more is not, and many wild species from Thailand, China, Brazil, Guatemala and elsewhere are at risk. One orchid species in ten is threatened and the continued loss of tropical forests means that many more will disappear before they are known to science. Even those on 'Orchis Bank', near Down House, survive only through the vigilance of local naturalists.

Britain itself, like many other places, faces a mirror image of the loss of its insect-eaters, orchids and more: a revenge of the immigrants, a wave of creatures that have appeared from almost nowhere and have attacked its natives. A New Zealand flatworm introduced to Belfast in the 1960s has run wild and an Australian cousin has also begun to move. It kills earthworms as it wraps itself around them and digests them alive. The pest has spread through Scotland, northern England and Ireland and in some places worm populations have collapsed.

The losers in the post-Victorian battle have been replaced by others that have thrived in the new global economy. Many have migrated to new places. There, they cause havoc. Weedy plants are a nuisance but weedy animals are even worse.

Modern society has always depended on aliens, creatures moved from their native lands, be they maize, chickens or cattle. They evolved in the Middle East, in Asia or in the New World, but have been transported to all parts of the globe. Many have become pests in their new home and many more have hitched a ride with those who cultivate them. Darwin himself noticed the invasion of British plants into the United States and asked his American colleague Asa Gray, 'Does it not hurt your Yankee pride that we thrash you so confoundedly?' The New World soon got its own back on the Old, with grey squirrels that eat woodland birds' eggs and Canadian pondweed that blocks streams. The third millennium is the era of the weeds and the weediest species of all – *Homo sapiens* – is to blame.

Plenty of weeds stay at home. They live in disturbed ground, flourish for a short time and move on to a new patch when conditions change. They do little damage except to the good temper of gardeners. When they escape, they are the botanical equivalents of pigs: they move in, exploit what is available and destroy the locals. Many imports from the Old World have thrived in the Americas. A common European roadside species, the knapweed, a small thistle with a pink or yellow flower, has covered tens of millions of hectares. It secretes a poison that kills native plants, which – unlike those at home – have not evolved resistance. As a sinister side-effect it also kills horses. The knapweed is now out of control. Others, such as the Brazilian water hyacinth, which has become a pest on Caribbean islands, find themselves in a place without their native pollinators and take up self-fertilisation. Some even hybridise with a relative, hijack its genes and gain renewed virulence as a result. The bright yellow and poisonous Oxford Ragwort common in disturbed ground in England is a hybrid between two Sicilian species brought to the Oxford Botanic Garden in the seventeenth century, which escaped and is still spreading.

Some of the most aggressive aliens are among the climbers. They are global pests. Even the hop has become a nuisance, with a Japanese variety that has spread across the United States. Kudzu, a climbing pea, is also native to Japan. In a gesture of amity, it was transplanted into that nation's ornamental garden at the 1876 Philadelphia Exposition. Gardeners liked the flowers and it was dispersed across the country. At first sight, the immigrant seemed helpful. It lays down roots two metres long and in the South was used to reduce soil loss after the forests had been cut down. The railroads gave free kudzu to farmers in the hope that they would cultivate it for fodder that their trucks could then transport. That was a considerable mistake. The weed grows so fast that the locals recommend, with an attempt at wit, that windows be closed at night to keep it out. In some places, it extends by thirty cen- timetres a day – twenty metres a season – and can soon smother a huge tree. Kudzu is out of control over an area that straddles Alabama, Georgia and Mississippi, and has spread as far north as Massachusetts and as far west as Texas. Attempts to subdue it cost half a billion dollars a year.

Other climbers are just as busy. Florida has 'air potatoes', yams from West Africa that sprawl over trees and block the light. It also suffers infestations of climbing ferns from Asia. English ivy has shaded out tracts of maple forest around Seattle. In Australia, the humble blackberry is a nuisance, as is the mile-a-minute vine, a morning glory introduced from the Old World tropics. Most are harmless at home, but a lifestyle that depends on a burst of growth when a sudden open space appears in the forest is lethal when exported to a place not adapted to their wiles.

Some of the climbers' success emerges from another by-prod- uct of human activity. The effects of the carbon crisis on climate are familiar enough – but it has unexpected side-effects, for climbers thrive in the new and enriched atmosphere. Over the past two decades, the proportion of the Amazon jungle taken up by lianas has gone up and up; in part because the forest has been opened up by loggers, but also because of the increased carbon dioxide, which they can soak up and lay down as wood. As a result they flourish at the expense of trees. Ivy, too, now grows at an exceptional rate as it gains extra carbon from the air.

Plenty more subjects of the Down House experiments have moved. Even barnacles are a menace. Some species, helped by the spread of shipping and by water (larvae included) dumped from ballast tanks, have begun to gallop across the globe. A barnacle the size of a tennis ball, once restricted to the Pacific coast of South America, has, within the past five years, made its way to Florida, Georgia and South Carolina. It infuriates boat owners because it acts as an unwelcome brake on their slippery bottoms. A species common on the coasts of California and Oregon has, in return, spread in huge numbers to Argentina. It was first recorded in the 1970s and has taken over the shoreline for two thousand kilometres. The animal has driven many local sea snails, seaweeds and more to extinction. It has found its way to Japan and at the present rate will soon reach Chile and wipe out the creature that drew the young naturalist's attention to the joys of the barnacle in the first place.

On land, too, a subterranean revolution is under way. Since Darwin's day the worm has turned. The animals he studied in his kitchen garden have crossed the world. The northernmost part of the Americas is almost devoid of native earthworms for they were wiped out by the last Ice Age, which left just a few remnants in the Pacific Northwest (one of which is still abundant on glaciers) and in a few patches elsewhere. Native Americans may deplore what the white man brought, but they did at least import earthworms, carried in pots or on the mud of immigrants' boots. Now fifty and more exotics have arrived. They can advance at several metres a year as they burrow and even faster when they hitch a ride. Escapes from anglers' bait tins mean that the creatures move at speed into remote forests as fishermen search for a new lake. Some are parthenogens and a single clone can take over huge tracts of land.

Unlike most invaders, they have been of some use. The West was won with the help of worms. Before the white man arrived, the northern prairies fed no more than herds of buffalo, but as the aliens spread soil fertility soared and maize and cattle moved in. Millions of hectares of soil were churned up and a dense and surly coat of acidic humus that sat on top of a sterile mineral layer was transformed into a well-mixed light soil with plenty of nutriments, just right for farmers.

Not all the news was good. Before the immigrants put in their appearance, many of the Northern states and large areas of Canada were covered by vast, fern-filled forests that sprang from deep mounds of leaf litter, or 'duff', that mouldered over years rather than being dragged into the ground by worms, as is the habit on the other side of the Atlantic. The duff was the home of beetles, salamanders, mice and more. The immigrants ate it, to leave a naked and unprotected surface. Most of the natives evolved to deal with undisturbed ground and suffer as a result. Thick undergrowth gives way to horsetails and pitcher plants. The local plants also depend on a relationship with the fungi that cluster around their roots – and they too have been lost under the assault. Aspen and birch trees die, native forests of sugar maple are parched as the water runs through the newly permeable soil, and prairie herbs disappear and are replaced by their European equivalents. In northern Minnesota, great tracts of hardwood have been destroyed in the past forty years and the problem is spreading. The once ponderous economy of those ancient forests has speeded up and vast quantities of carbon and nitrogen have been washed away. The danger is not limited to North America, for exotic worms have invaded tropical jungles. What they will do, we do not yet know.

Insects, too, followed the farmers and some, like worms, are a help. Most North American bees came from Europe. They arrived within two years of the Pilgrim Fathers for they were brought in by sweet-toothed pioneers anxious for honey. At once the immigrants set up wild colonies and thrived. So much were they an indication of European settlement that the Indians called them the 'white man's fly'. A wave of bees moved up the valley of the Missouri at fifty kilometres a year and, it was said, as the bee advanced, so the buffalo retreated. So impressed were the Mormons of Utah by the animal's hard work that they chose it as their state symbol.

European bees still live in their billions in North America, their numbers boosted by commercial hives. Like the worms, they have driven out local species, and – like them – they also play an important part in the agricultural economy. Many crops – fruit trees above all – need pollinators if they are to thrive. Eight out of ten of the top hundred or so food plants, responsible for two-thirds of global production, depend on them. California has so many almond, cherry and apple orchards that natural pollinators are unable to keep up and a healthy business in rented honey bees has grown up. It has an annual turnover of a hundred and fifty million dollars. A wild guess as to the overall value of bees to American farmers comes up with a figure a hundred times greater. As the season moves on, the hives follow the flowers. Without the added European bees, the yield of some fruit, seeds and nuts in the United States would drop by nine-tenths. The aliens may have damaged their local relatives but they have done a lot to help the humans who brought them.

The pollinators are in crisis and on both sides of the Atlantic the bees are in decline. The number of wild colonies of European bees in parts of North America is a tenth, and of those in hives a third, of what it was fifty years ago. Insecticides, parasites, viral disease and competition from introduced African bees have caused a crisis, as has the loss of hedgerows and other scrubby places that were a home for useful insects. In spite of attempts to keep the bees happy with wild flowers planted around orchids and by controlling insecticide use, the decline goes on. In an airborne twist to the tale, pollution kills the scent of many flowers and further reduces the chance that a pollinator will find its target. That is bad news for honey-lovers, worse for farmers and may be catastrophic for many native plants.

The troubles of the bees, the spread of the weeds and the destruction of the soil laid down by worms are just part of a new global crisis of agriculture. The population boom and the increased price of oil and fertilisers are also to blame. After a long period of stability, or even decline, the price of wheat, rice, chickpeas and other staples has gone up by several times in the years from 2006. The era of plenty may be near its end. India, for many years an exporter of food, has begun to ship in rice. The population explosion in Africa has led to shortage and while Chinese numbers are under better control, the nation's new affluence means a shift from cheap grains to expensive beef. World production of meat has gone up by four times since the 1960s. The habit is expensive in many ways. It takes fifty times more energy to make a kilogram of beef than a kilogram of maize or soy. Already there have been food riots in Haiti, Mexico and Egypt. The world's fisheries have been depleted and – in a fatuous gesture of ecological concern – some of its finest land used to grow biofuels. Soils are degraded and, as the climate changes, productive regions have lost their worth. Gloom is an occupational hazard for ecologists, but it is getting harder to be cheerful.

The fate of Charles Darwin's experimental subjects is part of a larger litany of decline. The Red List of Threatened Species has sixteen thousand names inscribed upon its pages and, every year, a spectacular creature – or several – is declared to be extinct. Many more pass almost unremarked and unmourned. If weeds and crops are to spread, others must pay the price. Most disappear as their homes are destroyed. Almost half the rain forest has gone, and mangrove swamps and Mediterranean landscapes face the same disaster with less publicity.

Extinction is part of evolution. Around half a million species of bird have lived since the group evolved more than a hundred

million years ago, but not many more than ten thousand lived at
a particular time – and that is the figure today (although twelve
hundred of them are threatened). Even so, nobody can deny that
we live in a time of rapid change, in which as some species thrive,
many more are doomed.

One small and specialised group of mammals – until not long
ago a vigorous part of the fauna of Europe, Africa and Asia – is
under particular threat. The globe, once filled with our hairy relatives, will soon be a Planet of the Ape. Almost half the world's six
hundred or so species of monkeys and apes face disaster. Habitat
destruction is the main threat, and hunters and disease also do a
lot to kill them off. Every large primate, except for one, is close
to the end of its evolutionary road. Just two hundred thousand
chimps are left in the wild. Gorillas have gone even faster and in
some places have died in multitudes from the Ebola virus, caught
from humans. Many populations of the orang-utan have been
broken up as the forest is destroyed and are already too small to
sustain themselves.

The genes show that the fortunes of the various great apes have
suffered a real reversal over the past several hundred thousand
years. The amount of variation hidden in DNA says a lot about
the abundance of any creatures in ancient times, for populations
that stay small for many generations lose genes through the
accidents of reproduction, while abundant animals can maintain
a pool of diversity that persists for long periods even if they
suffer a later collapse in numbers. The double helix hints that, for
most of that period, orangs, chimps and gorillas flourished while
Homo sapiens and his immediate ancestors struggled to survive.
Chimps, threatened as they now are, are three times more different at the molecular level, each from the next, than are humans.
Even within the past seventy thousand years, *Homo sapiens* went
through a bottleneck of just two or three thousand individuals
during a long age of drought. There have been plenty of local crises
since then as man filled new continents and set foot on remote

islands. For much of history we were an endangered primate while our relatives boomed. Now the shoe is on the other foot.

The world has six billion people and the figure will rise by half when the population peaks at about the time of *The Origin*'s bicentennial. *Homo sapiens* has, like the kudzu vine, the earthworm and the giant barnacle, begun to move and to multiply. As a result, its own future, like that of every other inhabitant of the planet, has been transformed. We may not have noticed it, but human evolution is in the middle of a shift as great as that of the species that surround us. We have long been the weediest primate of all and, in the past few years, have become far weedier than we were. History has always been made in bed, but the beds are closer together than they have ever been.

Evolution takes place in space as much as in time. As Darwin saw on the Galapagos – and as is manifest in the geography of human genes – populations isolated from each other diverge, either in response to local forces of natural selection or by random change. Other genetic trends have emerged over just the past few thousand years through the accidents of settlement, with reduced levels of variation in newly settled places such as the far Pacific, or even the New World as a whole. Ancient patterns, the remnants of the migration of peoples from the Levant to the British Isles, have been preserved only through the preference of most people to stay at home. That history will soon be lost.

How far was your birthplace from that of your partner and how far apart were your mother and father, and your grandmother and grandfather on each side, born? In almost every case, the distance has increased over the generations and continues to do so (my wife and I first saw the light five thousand kilometres apart, my mother and father about five; as my students say, it shows). Even in the Middle East, that centre of sexual conservatism, education, affluence and the chance to travel mean that DNA is on the move. A series of Israeli Arab villages experienced a drop in the incidence of cousin marriages from about a quarter in the 1980s to a sixth at the millennium. The same is true in Jordan, Lebanon and Palestine.

In the past, hurdles in the mind kept people apart. If the few remnants of hunter-gatherer tribes still left are any guide, any attempt to join another group was likely to be met by death. European frontiers, too, are marked by genetic steps, with a deeper difference in identity between the bluff beer-drinkers of the north and the vapid wine-bibbers of Spain and Italy. The continent's long history of staying at home is manifest in its surnames – those windows into sexual history – as much as its genes. A pedigree of names used to fit well with national boundaries with the Camerons more or less confined to Scotland and the Zapateros to Spain. Spain itself has the most localised patterns and can also boast of the commonest surname in Europe – García. In terms of names and genes, Paris is the most diverse city in Europe.

Names – and the DNA that accompanies them – are on the move. In 1881, Darwin's last full year, and the date of a British census, his surname was borne by about one Briton in fifty thousand. Its origin is Welsh, from 'derwen', or oak; a name transferred to the River Darwen in Lancashire. In his day its headquarters was in Sheffield and its surrounds, where the tag was eight times more common than in Britain as a whole. Almost all those who bore it lived within eighty kilometres of the city (the Joneses – the group with the second most common British surname after Smith – were still, in those happy days, more or less confined to Wales, where in some villages they formed a majority). By 1998, the naturalist's name had spread northwards to find a new centre in Durham, with secondary centres across the north from the Mersey to the Tyne and a minor outbreak in Herefordshire. The Joneses, too, had migrated, with a grand smear across the Welsh borders, north-west England and as far south as London. No longer must a Darwin or a Jones, or anyone else, marry – as so many once did – someone from their own family for lack of choice. Instead they come into contact with a diverse set of potential partners. The proportion of shared names in the marriage records of a typical English village has gone down by around 2 per cent a year even since the mid-1970s and by even more since the publication of *The Origin of Species*. Sheffield, once its author's nominal capital, now has scores of new names from across the world.

The United States has gone further down the road to homogeneity. Its telephone directories contain a million different surnames. Some remnants of history remain, with Wisconsin full of Scandinavians while the phone books of New Mexico, Colorado and Texas reveal the presence of many Spanish immigrants. Even so, the general picture is – unlike Europe – one of national uniformity. One in eight Americans is foreign-born (and in California twice as many), and Americans move home, on average, a dozen times in a lifetime, with ten million each year shifting from one state to another. Such frantic migration soon mixes up names, and, in time, genes. The power of the train or plane as an agent of evolution will soon even out yet more of man's genetic differences.

Almost everywhere, biological frontiers are becoming porous. An era of uniformity is at hand as vast hordes of people move in search of work or sunshine and, in the end, sex. In Britain, the proportion of those born abroad has doubled in the past fifty years and now represents a tenth of the population. Man, like the ecosystem he lives in, is in the midst of a grand averaging.

Intermarriage has been around for a long time. Many white Britons can trace a black ancestor from the small African population that lived in England several centuries ago. About half the men of a certain Yorkshire kindred, the family Revis, share a Y-chromosome type otherwise found only in West Africa. There have been Africans in Britain since the Romans and by the eighteenth century these islands held ten thousand black people. Since then the proportion of Britons who claim recent full or partial descent from Africa has gone up by twenty times and continues to climb. The popular view of the British as a nation walled into a series of ghettos is wrong. In 1991, one electoral ward in ten (a ward is the smallest parliamentary subdivision, with around eight thousand wards in England) had more than a tenth of its citizens from an ethnic minority. Ten years later, the figure was one in eight and by 2011 will rise to one in five. The growth comes in the main not from immigration, but from the movement of people within Britain and from the simple fact that young people, many of them migrants, have more children than do old.

All geneticists are firm believers in the healing power of lust; in the ability of desire to overcome social or geographic barriers. In 2001, about one British marriage in fifty – a quarter of a million in total, with many more couples cohabiting – was between part-ners from different ethnic groups. Hundreds of thousands of children have one parent from Britain and the other from the Caribbean and almost as many are the progeny of white and Asian parents. British Afro-Caribbean males are half again as likely to marry a white female than are black women to find a white hus-band, but for the Chinese those preferences are reversed. Such relationships are not, as often believed, found just among the poor for more than half of those involved live in the suburbs and are richer and more educated than average.

Assimilation is well under way in modern Britain, among the most sexually open nations in the world. In today's England, mate choice is made as much by level of education as by skin colour. Many other countries too, whether they like it or not, have opened up their gene pools. *Homo sapiens* – already the most tedious of all mammals in its geography – will soon, like the worms and the insects, be even more uniform than it was.

In that great global coalescence, *Homo sapiens* has evolved in just the same way as have other weeds. In other ways, man is quite unique: for he is the only animal that has escaped – or almost so – from the reach of evolution's pitiless laws of life and death. Natural

selection has long been at work on our species, even if our ingenuity has mitigated its power, with far less response in our own
line than in that of the chimpanzee. Now, in the nation in which
the idea was invented – and, more and more, in the world as a
whole – the process has slowed down and may soon stop.

I depress my first-year students with the statement that two out
of three of them will die for reasons connected to the genes they
carry (a vague claim, but good enough for undergraduates). Then
I try to cheer them up by pointing out that had I given the
lecture in Shakespeare's time, two out of three of them would,
at the age of eighteen or so, be dead already. Even in the year of
Darwin's birth, about half of all British newborns died before they
reached their majority. Life in these islands has seen a real change
for the better. An English baby born in the year of the millennium
had a 99 per cent chance of surviving until 2021 and that figure
continues to improve. Japan does even better, and the United States
rather worse, but most of the developed world has seen a revolution
in the prospects of the young. In most countries, even those as poor
as Ecuador, the death rate of children is less than, or no more than
twice as high as, the British figure. Africa is, alas, a real exception.
The death rate for under-fives is one in four in Sierra Leone and is
almost as high in Angola and Liberia, while Swaziland has an overall life expectancy of just forty years, half that of Japan. As a result,
differences in childhood mortality still account for most of the globe's
variation in life expectancy but, outside the continent where *Homo
sapiens* began, those differences have withered away.

Natural selection is very interested in the death of young
people, for they have not yet passed on their genes. It cares far less
about the fate of the old – those over forty or so – for their sexual
lives are over and their relevance to evolution at an end. As a
result, the great agent of change has lost a good part of its fuel. For
most of history, the Grim Reaper was the master of man's biological fate. Now, he is taking a rest, and inherited differences in the
ability to withstand cold, starvation, vitamin deficiency or disease

no longer do much to power the evolutionary machine. Plenty of people still die for those reasons, but they do so when they are elderly and evolution no longer notices them.

The Darwinian examination has two parts. In the developed world, most people pass the first paper: they stay alive until they are old enough to have children. The second section is just as inexorable but has a wider range of marks, for any candidate for evolutionary success has to find a mate and reproduce. The more children they have, the better the prospects for their genes.

Females are limited in the number of offspring they can produce by the mechanics of pregnancy and child care, while males are free to spread their sperm to a multitude of partners, even if a certain amount of persuasion is needed first. As a result, males compete for the attention of females, while females must decide which males should be allowed in. Sexual selection depends on the same logic as selection on the ability to survive: on inherited differences, not in the chances of life or death, but in the number of young. The rule applies to humans just as it does to birds and flowers.

Both humans and peacocks have more variation in male sexual success than in female. Until not long ago, many societies contained a few satisfied libertines, outnumbered by a huge mass of frustrated men. The powerful have always taken their amatory chances when they arise. Henry VIII was a minor player in the marital stakes. The emperor in the Ch'I dynasty of China maintained a palace with several thousand women available for his amusement, while in tribal societies men with high status still have many more mates than do their humble (and often celibate) fellows. Mohammed bin Laden, father of Osama of that ilk, had twenty-two wives and fifty-three children (and in the year of Osama's birth he had six). His best known son had, last time they were counted, five wives and twenty-two children. Given the equal numbers of men and women at birth, plenty of his henchmen will be obliged to die, naturally or otherwise, without issue.

The British Isles have a history of sexual inequality that makes even the antics of the founder of the Church of England look feeble. The evidence is in the Y chromosome, which marks male descent. It comes in a diversity of forms. In the majority of places most men have their own more or less unique model of their most masculine attribute. A fifth of the men of north-west Ireland, in contrast, share a more or less identical version of the Y chromosome – which means that they all trace ancestry from the same male.

In the glorious days of old Erin, sexual inequality was rife. Lord Turlough O'Donnell, who died in 1423, had eighteen sons and fifty-nine grandsons. He was himself a descendant of the High Kings of Ireland, all of whom claimed a certain fifth-century warlord, Niall of the Nine Hostages, a man who once kidnapped St Patrick, as their common ancestor. The Y chromosome of Niall the hostage-taker has, thanks to his own exploits and those of his powerful male descendants, spread to thousands of today's Irishmen. The surnames fit, too, for men of the Gallagher, O'Reilly and Quinn families, all of whom claim descent from the High Kings, are most likely to bear the special Y chromosome. In Ireland, for many centuries, the mightiest male passed on his genes, and many of his fellows passed their days in glum celibacy and the demand for soldiers or priests occupied their energies instead.

Today's reproductive universe is quite different. Everywhere, the weak and the powerful – the poor and the rich – are sexually closer than they were. For every class in society, the average number of children has, thanks to technology, gone down. Natural selection cares naught for that, for the important figure is not the number of progeny, but the variation in how many children people have. That figure has shrunk. Five centuries ago in Florence, the upper crust had twice as many offspring as did the peasantry but now the Florentine poor have more than the rich and Britain is much the same (which worries the *Daily Mail*).

The gulf has closed through restraint by the affluent rather than excess by the poor. A furtive exchange of information about birth control meant that rich families soon became smaller while those of the poor declined more slowly. Schools, too, are a powerful contraceptive. Everywhere, people with degrees have fewer children than those who drop out. As education spreads, the fertility imbalances will become smaller yet.

Inequality, in survival or in sex, is an ore refined by natural selection to adapt existence to changing conditions. The differences in people's ability to stay alive and to have children can be combined in a single measure that shows just how much of that raw material is still available. Across the world the figure – the 'opportunity for selection' – is in steep decline. India tells the tale within a single country. The nation encompasses a range of cultures from tribal hill-peoples to affluent urbanites, together with vast numbers of peasant farmers whose lives are rather like those of Europeans a few centuries ago. The figures of life, death and birth when put together show that natural selection has lost nine-tenths of its power in India's middle class when compared with the people of the tribes. The same is true when we compare the modern world with that of the Middle Ages and – to a lesser degree – even with that of the Victorians.

As Charles Darwin himself insisted, evolution is not a predictive science. Natural selection has no inbuilt tendency to improve matters (or, for that matter, to make them worse). For *Homo sapiens,* some nasty surprises no doubt lurk around the corner. Some day, evolution will take its revenge and we may fail in the struggle for existence against ourselves, the biggest ecological challenge of all.

Whatever the future holds, the bicentennial of his birth marks a new era in the biology of our planet. The changes are not limited to the rain forest, or the coral reefs, or the teeming tropics, but are hard at work on Darwin's own island and on the people who live there. From Shrewsbury to the Galapagos and from worms to barnacles to human beings, there has been a triumph of the average. The Earth is, as a result, a far less interesting place than it was when HMS *Beagle* set sail. Whether it becomes even less so – and whether it survives at all – depends on the talents of the only creature ever to step beyond the limits of Darwinian evolution.

INDEX